# Frequency-Domain Multiuser Detection for CDMA Systems

# RIVER PUBLISHERS SERIES IN COMMUNICATIONS

*Consulting Series Editors*

MARINA RUGGIERI
*University of Roma "Tor Vergata"*
*Italy*

HOMAYOUN NIKOOKAR
*Delft University of Technology*
*The Netherlands*

This series focuses on communications science and technology. This includes the theory and use of systems involving all terminals, computers, and information processors; wired and wireless networks; and network layouts, procontentsols, architectures, and implementations.

Furthermore, developments toward new market demands in systems, products, and technologies such as personal communications services, multimedia systems, enterprise networks, and optical communications systems.

- Wireless Communications
- Networks
- Security
- Antennas & Propagation
- Microwaves
- Software Defined Radio

For a list of other books in this series, please visit www.riverpublishers.com.

# Frequency-Domain Multiuser Detection for CDMA Systems

**Paulo Silva**
*DEE, ISE-UAlg, Faro, Portugal*

and

**Rui Dinis**
*DEE, FCT-UNL, Monte de Caparica, Portugal*

**River Publishers**

Routledge
Taylor & Francis Group

LONDON AND NEW YORK

**Published 2012 by River Publishers**
River Publishers
Alsbjergvej 10, 9260 Gistrup, Denmark
www.riverpublishers.com

**Distributed exclusively by Routledge**
4 Park Square, Milton Park, Abingdon, Oxon OX14 4RN
605 Third Avenue, New York, NY 10158

First published in paperback 2024

*Frequency-Domain Multiuser Detection for CDMA Systems* / by Paulo Silva, Rui Dinis.

*Routledge is an imprint of the Taylor & Francis Group, an informa business*

Publisher's Note
The publisher has gone to great lengths to ensure the quality of this reprint but points out that some imperfections in the original copies may be apparent.

While every effort is made to provide dependable information, the publisher, authors, and editors cannot be held responsible for any errors or omissions.

ISBN: 978-87-92329-70-7 (hbk)
ISBN: 978-87-7004-530-8 (pbk)
ISBN: 978-1-003-33820-8 (ebk)

DOI: 10.1201/9781003338208

# Contents

# List of Figures

# List of Acronyms

| | |
|---|---|
| 3G | Third Generation |
| 3GPP | Third Generation Partnership Project |
| A/D | Analog-to-Digital |
| AM | Amplitude Modulation |
| AWGN | Additive White Gaussian Noise |
| B3G | Beyond Third Generation |
| BER | Bit Error Rate |
| BLAST | Bell Laboratory Layered Space-Time |
| BS | Base Station |
| CDM | Code Division Multiplexing |
| CDMA | Coded Division Multiple Access |
| C&F | Clipping and Filtering |
| CIR | Channel Impulse Response |
| CP | Cyclic Prefix |
| D/A | Digital-to-Analog |
| DFE | Decision Feedback Equalizer |
| DFT | Discrete Fourier Transform |
| DS-CDMA | Direct Sequence-Coded Division Multiple Access |
| EGC | Equal Gain Combining |
| EXIT | Extrinsic Information Transfer |
| FDE | Frequency Domain Equalizer |
| FDM | Frequency Division Multiplexing |
| FDMA | Frequency Division Multiple Access |
| FT | Fourier Transform |
| FFT | Fast Fourier Transform |
| HD | Hard Decision |
| HIPERLAN | HIgh PERformance Local Area Network |
| i.i.d. | independent and identically distributed |
| IB-DFE | Iterative Block-Decision Feedback Equalizer |
| ICI | Inter-Channel Interference |

| | |
|---|---|
| IDFT | Inverse Discrete Fourier Transform |
| IF | Intermediate Frequency |
| IFFT | Inverse Fast Fourier Transform |
| IMP | Inter-Modulation Product |
| IMUD | Iterative Multiuser Detector |
| ISI | Inter-Symbol Interference |
| LLR | Log-Likelihood Ratio |
| LST | Layered Space-Time |
| MAI | Multiple Access Interference |
| MAP | maximum a posteriori |
| MC | Multicarrier |
| MC-CDMA | Multicarrier Coded Division Multiple Access |
| MFB | Matched Filter Bound |
| MIMO | Multiple-Input, Multiple Output |
| MLSE | Maximum Likelihood Sequence Estimator |
| MMSE | Minimum Mean-Squared Error |
| MRC | Maximal Ratio Combining |
| MSE | Mean-Squared Error |
| MSK | Minimum Shift Keying |
| MT | Mobile Terminal |
| MT-CDMA | Multitone-Coded Division Multiple Access |
| MUI | Multiuser Interference |
| MUD | Multiuser Detector |
| NL | Nonlinear |
| OA | Overlap-Add |
| OFDM | Orthogonal Frequency Division Multiplexing |
| OQPSK | Offset Quadrature Phase Shift Keying |
| OS | Overlap-Save |
| PA | Prefix Assisted |
| PAPR | Peak-to-Average Power Ratio |
| pdf | probability density function |
| PDP | Power Delay Profile |
| PIC | Parallel Interference Cancelation |
| PM | Phase Modulation |
| PMEPR | Peak-to-Mean Envelope Power Ratio |
| PN | Pseudo Noise |
| PSD | Power Spectral Density |
| PSK | Phase Shift Keying |
| QAM | Quadrature Amplitude Modulation |

| | |
|---|---|
| QPSK | Quadrature Phase Shift Keying |
| RF | Radio Frequency |
| SC | Single Carrier |
| SC-FDE | Single Carrier-Frequency Domain Equalizer |
| SD | Soft Decision |
| SDMA | Space Division Multiple Access |
| SIC | Successive Interference Cancelation |
| SINR | Signal-to-Interference plus Noise Ratio |
| SISO | Soft In, Soft Out |
| SIR | Signal-to-Interference Ratio |
| SNR | Signal-to-Noise Ratio |
| SS | Spread Spectrum |
| SSPA | Solid State Power Amplifier |
| SU | Single User |
| TDMA | Time Division Multiple Access |
| TWTA | Traveling Wave Tube Amplifier |
| UMTS | Universal Mobile Telecommunication System |
| UW | Unique Word |
| ZF | Zero Forcing |
| ZP | Zero Padded |

# List of Symbols

## General Symbols

| | |
|---|---|
| $A_m$ | $m$th frequency-domain data symbol |
| $A_{m,p}$ | $m$th frequency-domain data symbol of the $p$th user |
| $A_{m,p}^I$ | real part of $A_{m,p}$ ("in-phase bit") |
| $A_{m,p}^Q$ | imaginary part of $A_{m,p}$ ("quadrature bit") |
| $\tilde{A}_{m,p}$ | data estimate of $A_{m,p}$ |
| $\hat{A}_{m,p}$ | "hard decision" of $A_{m,p}$ |
| $\overline{A}_{m,p}$ | "soft decision" of $A_{m,p}$ |
| $A_{k,l}^{(p)}$ | $k$th frequency-domain data symbol transmitted by the $l$th antenna of the $p$th user |
| $\tilde{A}_{k,l}^{(p)}$ | data estimate of $A_{k,l}^{(p)}$ |
| $\hat{A}_{k,l}^{(p)}$ | "hard decision" of $A_{k,l}^{(p)}$ |
| $\overline{A}_{k,l}^{(p)}$ | "soft decision" of $A_{k,l}^{(p)}$ |
| $a_m$ | $m$th time-domain data symbol |
| $a_{m,p}$ | $m$th time-domain data symbol of the $p$th user |
| $a_{m,p}^I$ | real part of $a_{m,p}$ ("in-phase bit") |
| $a_{m,p}^Q$ | imaginary part of $a_{m,p}$ ("quadrature bit") |
| $\tilde{a}_{m,p}$ | data estimate of $a_{m,p}$ |
| $\hat{a}_{m,p}$ | "hard decision" of $a_{m,p}$ |
| $\overline{a}_{m,p}$ | "soft decision" of $a_{m,p}$ |
| $a_{m,l}^{(p)}$ | $m$th time-domain data symbol transmitted by the $l$th antenna of the $p$th user |
| $\tilde{a}_{m,l}^{(p)}$ | data estimate of $a_{m,l}^{(p)}$ |
| $\hat{a}_{m,l}^{(p)}$ | "hard decision" of $a_{m,l}^{(p)}$ |
| $\overline{a}_{m,l}^{(p)}$ | "soft decision" of $a_{m,l}^{(p)}$ |
| $B_k$ | feedback equalizer coefficient for the $k$th frequency |
| $B_{k,p}$ | feedback equalizer coefficient for the $k$th frequency and the $p$th user |

$B_{k,p}^{(p')}$ feedback equalizer coefficient for the $k$th frequency, $p$th user and the $p'$th interferer

$B_{k,l,l'}^{(p,p')}$ feedforward equalizer coefficient for the $k$th frequency, $l$th transmit antenna of the $p$th user and $l'$th transmit antenna of the $p'$th interferer

$C_H$ high power users class

$C_L$ low power users class

$C_k$ $k$th chip of the frequency-domain spreading sequence

$C_{k,p}$ $k$th chip of the $p$th frequency-domain spreading sequence

$C_{k,l}^{(p)}$ $k$th chip transmitted by the $l$th antenna of the $p$th frequency-domain spreading sequence

$c_n$ $n$th chip of the time-domain spreading sequence

$c_{n,p}$ $n$th chip of the $p$th time-domain spreading sequence

$c_{n,l}^{(p)}$ $n$th chip transmitted by the $l$th antenna of the $p$th time-domain spreading sequence

$D_k$ $k$th frequency-domain nonlinear distortion component

$D_k^{Tx}$ transmitted $k$th frequency-domain nonlinear distortion component

$D_{k,p}$ $k$th frequency-domain nonlinear distortion component of the $p$th user

$\hat{D}_{k,p}$ "hard estimate" of $D_{k,p}$

$\overline{D}_{k,p}$ "soft estimate" of $D_{k,p}$

$D_{k,p}^{Res}$ $k$th frequency-domain residual nonlinear distortion component of the $p$th user

$\hat{D}_{k,p}^{Res}$ "hard estimate" of $D_{k,p}^{Res}$

$\overline{D}_{k,p}^{Res}$ "soft estimate" of $D_{k,p}^{Res}$

$d_n$ $n$th time-domain nonlinear distortion component

$d_m^{eq}$ equivalent $m$th time-domain nonlinear distortion component for detection proposes

$d_n^{Tx}$ $n$th time-domain transmitted nonlinear distortion component

$d_{n,p}$ $n$th time-domain nonlinear distortion component of the $p$th user

$E_b$ average bit energy

$F$ subcarrier separation

$F_k$ feedforward equalizer coefficient for the $k$th frequency

$F_{k,p}$ feedforward equalizer coefficient for the $k$th frequency and $p$th user

$F_k^{(l)}$ feedforward equalizer coefficient for the $k$th frequency and $l$th diversity branch

| | |
|---|---|
| $F_{k,p}^{(l)}$ | feedforward equalizer coefficient for the $k$th frequency, $l$th diversity branch and $p$th user |
| $F_{k,l}^{(p,r)}$ | feedforward equalizer coefficient for the $k$th frequency, $l$th transmit antenna of the $p$th user at $r$th receive branch |
| $f$ | frequency variable |
| $f_k$ | $k$th frequency |
| $G_d(k)$ | Power spectrum of $d_n$ (DFT of the autocorrelation $R_d(n)$) |
| $G_s(k)$ | Power spectrum of $s_n$ (DFT of the autocorrelation $R_s(n)$) |
| $G_s^C(k)$ | Power spectrum of $s_n^C$ (DFT of the autocorrelation $R_s^C(n)$) |
| $G_k$ | frequency-domain filtering coefficient for the $k$th frequency |
| $g_C(x)$ | nonlinear envelope clipping characteristic |
| $H_k$ | overall channel frequency response for the $k$th frequency |
| $H_{k,p}$ | overall channel frequency response for the $k$th frequency and $p$th user |
| $H_{k,p}^{Ch}$ | channel frequency response for the $k$th frequency and $p$th user |
| $H_k^{(l)}$ | overall channel frequency response for the $k$th frequency and $l$th diversity branch |
| $H_{k,p}^{(l)}$ | overall channel frequency response for the $k$th frequency, $l$th diversity branch and $p$th user |
| $H_{k,l}^{(p,r)}$ | overall channel frequency response for the $k$th frequency transmitted by the $l$th antenna of the $p$th user at the $r$th receive diversity branch |
| $H_{k,l}^{Ch,p,r}$ | channel frequency response for the $k$th frequency transmitted by the $l$th antenna of the $p$th user at the $r$th receive diversity branch |
| $H_T(t)$ | frequency response of the reconstruction filter |
| $h_T(t)$ | impulsive response of the reconstruction filter |
| $J$ | Lagrange function |
| $K$ | spreading factor |
| $K_r$ | spreading factor associated to the $r$th resolution |
| $\mathcal{K}_F$ | normalization constant for the FDE |
| $k$ | frequency index |
| $L$ | space diversity order |
| $L_R$ | number of receive antennas |
| $L_T$ | number of transmit antennas |
| $L_T^{(p)}$ | number of transmit antennas of the $p$th user |
| $L_{m,p}$ | log-likelihood ratio for the $m$th symbol and $p$th user |
| $L_{m,p}^I$ | log-likelihood ratio of the "in-phase bit" for the $m$th symbol and $p$th user |

| | |
|---|---|
| $L_{m,p}^Q$ | log-likelihood ratio of the "quadrature bit" for the $m$th symbol and $p$th user |
| $L_\gamma^{(1)}(x)$ | generalized Laguerre polynomial of order $\gamma$ |
| $l$ | antenna index |
| $M$ | number of data symbols per block for each spreading code/user |
| $M_{Tx}$ | oversampling factor |
| $m$ | data symbol index |
| $N$ | number of samples/subcarriers |
| $N_0$ | noise power spectral density (unilateral) |
| $N_k$ | channel noise for the $k$th frequency |
| $N_k^{(l)}$ | channel noise for the $k$th frequency and $l$th diversity branch |
| $N_G$ | number of guard samples |
| $N_I$ | number of idle subcarriers |
| $N_L$ | number of transmitted layers |
| $n$ | time index |
| $P$ | number of users |
| $P_N^{eq}$ | noise power associated to $w_n^{eq}$ |
| $P_{NL}$ | total power at the nonlinearity output |
| $P_{NL}^S$ | power of the useful component at the nonlinearity output |
| $P_{NL}^I$ | power of the self-interference component at the nonlinearity output |
| $P_{Tx}^S$ | transmitted power of the useful component |
| $P_{Tx}^I$ | transmitted power of the self-interference component |
| $P_{Tx,r'}^S$ | transmitted power of the useful component for the $r'$th spreading sequence |
| $P_{Tx,IB}^I$ | transmitted in-band power of the self-interference component |
| $P_b$ | AWGN channel performance |
| $P_{b,MFB}$ | matched filter bound performance |
| $P_{b,Ray}$ | Rayleigh fading channel performance |
| $P_{b,SU}$ | single user performance |
| $P_{e,p}$ | bit error rate for the $p$th user |
| $P_{2\gamma+1}$ | total power associated to the IMP of order $2\gamma + 1$ |
| $p$ | user index |
| $p_X(x)$ | probability density funtion |
| $Q(x)$ | Q function |
| $R(f)$ | Fourier transform of $r(t)$ |
| $R_d(n)$ | autocorrelation of $d_{n,p}$ |

| | |
|---|---|
| $R_s(n)$ | autocorrelation of $s_{n,p}$ |
| $R_s^C(n)$ | autocorrelation of $s_{n,p}^C$ |
| $r$ | resolution index |
| $r(t)$ | rectangular impulse; shaping impulse |
| $S(f)$ | frequency-domain signal |
| $S_k$ | $k$th frequency-domain data symbol/chip |
| $S_k^{(m)}$ | $k$th frequency-domain data symbol of the $m$th block |
| $\tilde{S}_k$ | estimate for the $k$th frequency-domain data symbol/chip |
| $\hat{S}_k$ | "hard decision" for the $k$th frequency-domain data symbol/chip |
| $\overline{S}_k$ | "soft decision" for the $k$th frequency-domain data symbol/chip |
| $S_k^C$ | $k$th frequency-domain chip with clipping |
| $S_k^{CF}$ | $k$th frequency-domain chip with clipping and filtering |
| $S_{k,p}$ | $k$th frequency-domain chip of the $p$th user |
| $\tilde{S}_{k,p}$ | estimate for the $k$th frequency-domain chip of the $p$th user |
| $\hat{S}_{k,p}$ | "hard decision" for the $k$th frequency-domain chip of the $p$th user |
| $\overline{S}_{k,p}$ | "soft decision" for the $k$th frequency-domain chip of the $p$th user |
| $S_{k,l}^{(p)}$ | $k$th frequency-domain chip transmitted by the $l$th antenna of the $p$th user |
| $\tilde{S}_{k,l}^{(p)}$ | sample estimate of $S_{k,l}^{(p)}$ |
| $\hat{S}_{k,l}^{(p)}$ | "hard decision" of $S_{k,l}^{(p)}$ |
| $\overline{S}_{k,l}^{(p)}$ | "soft decision" of $S_{k,l}^{(p)}$ |
| $S_{k,p}^{Tx}$ | transmitted $k$th frequency-domain chip of the $p$th user |
| $\hat{S}_{k,p}^{Tx}$ | "hard decision" of $S_{k,p}^{Tx}$ |
| $S_{k,p}^C$ | $k$th frequency-domain chip of the $p$th user with clipping |
| $S_{k,p}^{CF}$ | $k$th frequency-domain chip of the $p$th user with clipping and filtering |
| $s(t)$ | time-domain signal |
| $s^{(m)}(t)$ | signal associated to the $m$th data block |
| $s^{(m,P)}(t)$ | periodic signal associated to the $m$th block |
| $s^{Tx}(t)$ | transmitted signal |
| $s_M$ | clipping level |
| $s_n$ | $n$th time-domain data symbol/chip |
| $\tilde{s}_n$ | sample estimate for the $n$th time-domain data symbol/chip |
| $\hat{s}_n$ | "hard decision" of the $n$th time-domain data symbol/chip |
| $\overline{s}_n$ | "soft decision" of the $n$th time-domain data symbol/chip |
| $s_n^C$ | $n$th time-domain chip with clipping |
| $s_n^{Tx}$ | transmitted $n$th time-domain chip |

| | |
|---|---|
| $s_n^U$ | "useful" component of the transmitted $n$th time-domain chip |
| $s_n^{CF}$ | $n$th time-domain chip with clipping and filtering |
| $s_{n,p}$ | $n$th time-domain chip of the $p$th user |
| $\tilde{s}_{n,p}$ | sample estimate of $s_{n,p}$ |
| $\hat{s}_{n,p}$ | "hard decision" of $s_{n,p}$ |
| $\overline{s}_{n,p}$ | "soft decision" of $s_{n,p}$ |
| $s_{n,p}^C$ | $n$th time-domain chip of the $p$th user with clipping |
| $s_{n,p}^{CF}$ | $n$th time-domain chip of the $p$th user with clipping and filtering |
| $s_{n,p}^{Tx}$ | transmitted $n$th time-domain chip of the $p$th user |
| $s_{n,l}^{(p)}$ | $n$th time-domain chip transmitted by the $l$th antenna of the $p$th user |
| $\tilde{s}_{n,l}^{(p)}$ | sample estimate of $s_{n,l}^{(p)}$ |
| $\hat{s}_{n,l}^{(p)}$ | "hard decision" of $s_{n,l}^{(p)}$ |
| $\overline{s}_{n,l}^{(p)}$ | "soft decision" of $s_{n,l}^{(p)}$ |
| $T$ | duration of the useful part of the block |
| $T_B$ | block duration |
| $T_G$ | guard period |
| $T_c$ | chip duration |
| $T_s$ | symbol duration |
| $t$ | time variable |
| $Y_k$ | received sample for the $k$th frequency |
| $Y_k^{(l)}$ | received sample for the $k$th frequency and $l$th diversity branch |
| $Y_k^{Corr}$ | corrected received sample for the $k$th frequency |
| $Y_k^{Corr(l)}$ | corrected received sample for the $k$th frequency and $l$th diversity branch |
| $y_n$ | $n$th time-domain received sample |
| $y_n^{(l)}$ | $n$th time-domain received sample for the $l$th diversity branch |
| $y(t)$ | received signal |
| $w_n$ | $n$th channel noise sample |
| $w_m^{eq}$ | equivalent $m$th channel noise sample for detection proposes |
| $W_{2\gamma+1}(x)$ | auxiliary function for the IMP power computation |

## Greek Letters Symbols

| | |
|---|---|
| $\Delta_k$ | error term for the $k$th frequency-domain "hard decision" estimate |
| $\Delta_{k,p}$ | error term for the $k$th frequency-domain "hard decision" estimate |

|  |  |
|---|---|
|  | of the $p$th user |
| $\Delta_{k,l}^{(p)}$ | error term for the $k$th frequency-domain "hard decision" estimate of the $l$th transmit antenna of the $p$th user |
| $\Psi_m$ | set of frequencies employed to transmit the $m$th data symbol |
| $\Psi_m$ | set of indexes $r'$ of the spreading codes associated to the $r$th resolution |
| $\alpha$ | scale factor associated to the nonlinear operation |
| $\alpha_k$ | frequency shaping associated to the iterative C&F procedure for the $k$th frequency |
| $\alpha_k^{Tx}$ | frequency shaping associated to the iterative C&F procedure for the transmitted $k$th frequency |
| $\alpha_{k,p}$ | frequency shaping associated to the iterative C&F procedure for the $k$th frequency of the $p$th user |
| $\beta$ | inverse of the signal-to-noise ratio |
| $\gamma_p$ | average channel frequency response for the $p$th user |
| $\varepsilon_n^{eq}$ | overall noise for the $n$th time-domain symbol |
| $\varepsilon_K^{Eq}$ | overall error for the $k$th frequency-domain symbol |
| $\lambda$ | Lagrange multiplier |
| $v_{m,p}^I$ | error coefficient for the "in phase bit" of the $m$th symbol of the $p$th user |
| $v_{m,p}^Q$ | error coefficient for the "quadrature bit" of the $m$th symbol of the $p$th user |
| $\xi_p$ | power control weighting coefficient for the $p$th user |
| $\xi_r$ | power control weighting coefficient for the $r$th resolution |
| $\xi_l^{(p)}$ | power control weighting coefficient for the $p$th user and $l$th transmit antenna |
| $\eta_k$ | nonlinear distortion-to-signal ratio for the $k$th frequency |
| $\eta_S$ | degradation factor associated to the self-interference component |
| $\eta_{MF}$ | degradation factor associated to the filtering effects of $\alpha_k^{Tx}$ |
| $\rho_p$ | correlation coefficient of the $p$th user |
| $\rho_p^{(i)}$ | correlation coefficient of the $p$th user at the $i$th iteration |
| $\rho_{m,p}^I$ | correlation coefficient of the "in-phase bit" for the $m$th data symbol of the $p$th user |
| $\rho_{m,p}^Q$ | correlation coefficient of the "quadrature bit" for the $m$th data symbol of the $p$th user |
| $\sigma_A^2$ | variance of the transmitted frequency-domain data symbols |
| $\sigma_N^2$ | variance of the channel noise |
| $\sigma_p^2$ | variance of the data symbols estimates of the $p$th user |

$\sigma_S^2$      variance of the transmitted frequency-domain data symbols

$\sigma_{S,p}^2$      variance of the transmitted data symbols of the $p$th user

$\sigma_{D,p}^2$      variance of the frequency-domain nonlinear distortion of the $p$th user

$\sigma_s^2$      variance of the transmitted time-domain data symbols

$\Theta_k$      overall error for the $k$th frequency-domain sample

$\Theta_{m,p}$      overall error for the $m$th frequency-domain symbol of the $p$th user

$\theta_n$      overall error for the $n$th time-domain sample

$\theta_{m,p}$      overall error for the $m$th time-domain symbol of the $p$th user

$\tau_i$      delay associated to the $i$th path

$\nabla$      nabla operator

## Matrix Symbols

$\mathbf{A}(k)$      data symbol vector

$\tilde{\mathbf{A}}(k)$      data symbol estimate vector

$\hat{\mathbf{A}}(k)$      "hard decisions" data symbol vector

$\overline{\mathbf{A}}(k)$      "soft decisions" data symbol vector

$\mathbf{B}(k)$      feedback equalizer coefficients matrix

$\mathbf{B}^{(p)}(k)$      feedback equalizer coefficients vector for the $p$th interferer

$\mathbf{B}_p(k)$      feedback equalizer coefficients vector for the $p$th user

$\mathbf{B}_{k,l}^{(p)}$      feedback equalizer coefficients vector for the $l$th transmit antenna of the $p$th user

$\mathbf{D}(k)$      nonlinear distortion matrix

$\mathbf{D}^{Tx}(k)$      transmitted nonlinear distortion vector

$\hat{\mathbf{D}}(k)$      "hard estimate" of $\mathbf{D}(k)$

$\overline{\mathbf{D}}(k)$      "soft estimate" of $\mathbf{D}(k)$

$\mathbf{D}^{Res}(k)$      residual nonlinear distortion matrix

$\hat{\mathbf{D}}^{Tx}(k)$      estimate of $\mathbf{D}^{Tx}(k)$

$\mathbf{F}(k)$      feedforward equalizer coefficients matrix

$\mathbf{F}^{(l)}(k)$      feedforward equalizer coefficients matrix for the $l$th diversity branch

$\mathbf{F}_p(k)$      feedforward equalizer coefficients vector for the $p$th user

$\mathbf{F}_{k,l}^{(p)}$      feedforward equalizer coefficients vector for the $l$th transmit

| | |
|---|---|
| | antenna of the $p$th user |
| $\mathbf{H}(k)$ | channel frequency response matrix |
| $\mathbf{H}^{Ch}(k)$ | channel frequency response matrix for nonlinear transmitters |
| $\mathbf{H}^{Use}(k)$ | useful channel frequency response matrix for nonlinear transmitters |
| $\mathbf{H}^{(l)}(k)$ | channel frequency response matrix for the $l$th diversity branch |
| $\mathbf{H}_p(k)$ | channel frequency response vector for the $p$th user |
| $\mathbf{I}_N$ | $N \times N$ identity matrix |
| $\mathbf{J}$ | Lagrange matrix |
| $\mathbf{Y}(k)$ | received samples vector |
| $\mathbf{Y}^{Corr}(k)$ | corrected received samples vector |
| $\mathbf{Y}^{(l)}(k)$ | received samples vector for the $l$th diversity branch |
| $\mathbf{N}(k)$ | channel frequency noise vector |
| $\mathbf{N}^{(l)}(k)$ | channel frequency noise vector for the $l$th diversity branch |
| $\mathbf{P}$ | correlation coefficients matrix |
| $\mathbf{P}^{(p)}$ | correlation coefficients matrix for the $p$th user |
| $\mathbf{Q}$ | constant normalization matrix |
| $\mathbf{Q}_p$ | constant normalization vector for the $p$th user |
| $\mathbf{R}_A$ | autocorrelation matrix of $\mathbf{A}(k)$ |
| $\mathbf{R}_D$ | autocorrelation matrix of $\mathbf{D}(k)$ |
| $\mathbf{R}_N$ | autocorrelation matrix of $\mathbf{N}(k)$ |
| $\mathbf{R}_\Delta$ | autocorrelation matrix of $\mathbf{\Delta}(k)$ |
| $\mathbf{U}(k)$ | diagonal matrix with the fraction of the useful signal component |
| $\mathbf{V}(k)$ | equivalent matrix for the feedforward coefficients matrix equation |
| $\mathbf{\Delta}(k)$ | error terms matrix |
| $\mathbf{\Gamma}(k)$ | average channel frequency response matrix at the FDE output |
| $\mathbf{\Gamma}_p(k)$ | average channel frequency response vector for the $p$th user at the FDE output |
| $\mathbf{\Theta}(k)$ | overall symbols error vector |
| $\mathbf{\Lambda}$ | matrix with the Lagrange coefficients |
| $\mathbf{\Lambda}_p$ | vector with the Lagrange coefficient for the $p$th user |

## Common Operators

| | |
|---|---|
| $\lfloor x \rfloor$ | floor function (larger integer less then or equal to $x$) |
| $x \bmod y$ | modulus operator (reminder of the division of $x$ by $y$) |

| | |
|---|---|
| $\mathbf{A}^T$ | transpose of matrix $\mathbf{A}$ |
| $\mathbf{A}^*$ | conjugate of matrix $\mathbf{A}$ |
| $\mathbf{A}^H$ | hermitian of matrix $\mathbf{A}$ |
| diag($\mathbf{A}$) | diagonal of matrix $\mathbf{A}$ |
| tr($\mathbf{A}$) | trace of matrix $\mathbf{A}$ |
| $\delta_{n,m}$ | Kronecker delta function |

# 1

# Introduction

## 1.1 Motivation and Scope

The design of future broadband wireless systems presents a big challenge, since these systems should be able to cope with severely time-dispersive channels and are expected to provide a wide range of services (which may involve data rates of several tens of Mbit/s) and to have high spectral and power efficiencies. The use of equalization techniques to deal with the channel time-dispersion effects associated to the multipath signal propagation between transmitter and receiver becomes inevitable to compensate the inherent signal distortion levels and ensure good performance. However, implementation complexity and power consumption cannot be too high, especially at the mobile terminals (MT), since low-cost and relatively long live batteries are desirable at these terminals. The optimum receiver structure for time-dispersive channels corresponds to the well-known Viterbi equalizer [1], whose complexity grows exponentially with the channel impulse response length, making it recommendable only for channels whose impulse response spans over just a few symbols.

Time-domain equalization techniques are traditionally employed to mitigate channel frequency selectivity effects, leading to a much lower implementation complexity than Viterbi equalizers. Nonlinear equalizers such as decisions feedback equalizers (DFE) [2] can significantly outperform linear equalizers and have a good complexity/performance tradeoff. However, for conventional single carrier (SC) modulations the signal processing complexity (number of arithmetic operations per data symbol) required to mitigate the strong intersymbol interference (ISI) levels inherent to digital transmission over severely time-dispersive channels rapidly becomes prohibitive, especially when conventional time-domain equalization is employed at the receiver side.

1

SC transmission schemes have been the traditional digital communications format since the early days of telegraphy to many of the earlier wireless communications systems. In recent years, multicarrier (MC) modulations schemes combined with frequency-domain receivers implementations, especially the ones belonging to the orthogonal frequency division multiplexing (OFDM) class [3–5], became popular and are widely used in several broadband wireless communication systems which have to deal with strongly frequency-selective fading channels. The main reasons for the success of OFDM schemes is their robustness against time dispersion effects inherent to high transmission rate over severely time-dispersive channels together with low implementation complexity based on frequency-domain processing of signals blocks, using discrete Fourier transform (DFT) implemented by fast Fourier transform (FFT) [6]. Generation and block processing of signals in the frequency-domain, is enormously simplified by FFTs, yielding a signal processing complexity that grows logarithmically with the channel impulsive response (CIR) length.

Conventional SC modulations have also been shown to be suitable for block transmission schemes which provide a low implementation complexity: as with current OFDM-based schemes, each SC-based transmitted block includes a cyclic-prefix (CP), long enough to cope with the maximum relative channel delay, so that a linear frequency-domain equalization (FDE) technique, involving simple FFT computations, can be used at the receiver side [7]. These low-complexity single carrier frequency-domain equalization (SC-FDE) schemes are closely related to the ones employed in CP-assisted MC schemes (e.g., OFDM) and known to be derived from MC transmission ones, by shifting an inverse discrete Fourier transform (IDFT) from the transmitter to the receiver.

Spread spectrum (SS) techniques, particularly coded division multiple access (CDMA) techniques have been used in many communications and navigation systems over the past decades. CDMA schemes are also recognized as an appropriate multiple access scheme. In fact, when compared with conventional access techniques such as time-division multiple access (TDMA) and frequency-division multiple access (FDMA), CDMA schemes provide higher capacity and flexibility as well as some robustness to the hostile channel frequency-selectivity. Moreover, CDMA techniques can be combined with CP-assisted block transmissions, allowing frequency-domain receiver design with relatively low complexity, even for severely time-dispersive channels. The most popular CDMA techniques are DS-CDMA (Direct Sequence-CDMA) [8] and MC-CDMA (Multicarrier-CDMA) [9] techniques, which can combine a CDMA scheme with an SC-FDE or OFDM modulation, respectively. However, the interference levels caused by simultaneous transmitting users can be very high for both downlink (from the base station (BS) to the MT) and uplink transmissions (from the MT to the BS), particularly in the second case, since orthogonality between spreading codes associated to the different users is lost due to the channel effect. Therefore, significant performance degradation can be expected, especially for severely time-dispersive channels, when the number of users is high and/or when different users have different assigned powers. In these cases, more sophisticated receivers able to cope with these effects and improve the performance to acceptable levels need to be employed.

This book presents an unified study, relying on a CP-assisted block transmission approach, of enhanced frequency-domain multiuser detection (MUD) techniques with iterative signal detection/decoding techniques, in the context of DS-CDMA and MC-CDMA systems. Suitable transmit/receive structures with different complexity/performance tradeoffs are developed and analyzed for severely time-dispersive channels where high levels of interference are experienced, especially when the number of users is high and/or when different users have different assigned powers. Approaches for improvement of both the downlink and uplink transmissions will be treated, including multi-antenna techniques for performance and/or capacity enhancement. Appropriate multiple-input, multiple-output (MIMO) techniques for the uplink transmission are considered, involving sophisticated processing only at the receiver side. A special attention is devoted to iterative detection methods based on cancelation of interferences using the iterative block decision feedback equalization concept (IB-DFE) [10], with or without involving the decoding process. These methods are also combined with iterative estimation and threshold-based cancelation of deliberate nonlinear distortion effects introduced at the transmitter (to reduce the envelope fluctuations and peak-to-mean envelope power ratio (PMEPR) of the transmitted signals).

## 1.2 Outline

After this first introductory chapter, Chapter 2 is dedicated to the basic principles of MC modulations and their relations with SC modulations. OFDM modulations and SC modulations with linear FDE and with decision feedback equalization are also described, including the characterization of the transmitted signals in the time and frequency domains and the transmitter and receiver structures. The concepts of MC and SC modulations are then extended to SS techniques, particularly the CDMA schemes MC-CDMA and DS-CDMA with frequency-domain spreading, for both downlink and uplink transmissions in multiuser scenarios.

Chapter 3 deals with the receiver design for the downlink transmission of CP-assisted DS-CDMA and MC-CDMA systems by considering the use of IB-DFE techniques in space diversity receivers, as an alternative to conventional linear FDE techniques. The IB-DFE receiver parameters are derived and the turbo equalization concept based on IB-DFE receivers is defined by making use of "soft decisions" from the channel decoder outputs, in the feedback loop of the equalizer. Finally, a set of performance results is presented and discussed.

Chapter 4 is dedicated to the receiver design for the uplink transmission of DS-CDMA and MC-CDMA systems. Iterative frequency-domain MUD receivers, combining turbo equalization and multiple access interference (MAI) cancelation techniques, are proposed for both DS-CDMA and MC-CDMA systems. Parallel and successive interference cancelation (PIC and SIC) receiver structures are considered. An extension for MIMO systems is also presented. System implementation and signal processing complexity issues concerning the proposed iterative MUD receivers are also addressed and a set of performance results is presented and discussed.

Chapter 5 is concerned with the use of nonlinear transmitters in both DS-CDMA and MC-CDMA systems so as to reduce the envelope fluctuations of the transmitted signals without compromising the spectral efficiency. We consider suitable signal processing techniques involving a nonlinear time-domain operation followed by a frequency-domain filtering operation. The use of these techniques reduces the linearity requirements for the power amplifiers and/or the required back-off, increasing the power efficiency. The statistical characterization of the transmitted signals, which takes advantage of the Gaussian nature of the signals, is made and used for performance evaluation purposes. This chapter also includes enhanced receiver structures for both systems combining signal detection with iterative estimation and cancelation of deliberate nonlinear distortion effects introduced at the transmitter and the derivation of the receiver parameters and the nonlinear distortion estimates used to compensate for nonlinear effects. Performance results for the proposed DS-CDMA and MC-CDMA receiver structures are also presented and discussed.

Finally, Chapter 6 presents the final conclusions of this work and outlines some future work perspectives.

## 1.3 Notation and Simulations Aspects

Throughout this book, the following notation is adopted. Bold upper case letters denote matrices or vectors; $\mathbf{I}_N$ denotes the $N \times N$ identity matrix; $\mathbf{0}_{N \times M}$ denotes the $N \times M$ zero matrix; $(\cdot)^T$, $(\cdot)^H$, $(\cdot)^*$ and diag$(\cdot)$ denote the transpose, hermitian, conjugate and diagonal matrices, respectively; tr$(\mathbf{X})$ denotes the trace of matrix $\mathbf{X}$; $[\mathbf{X}]_{n,m}$ denotes the element of line $n$ and column $m$ of matrix $\mathbf{X}$. In general, lower case letters denote time-domain variables and upper case letters denote frequency-domain letters; $(\tilde{\cdot})$, $(\hat{\cdot})$ and $\overline{(\cdot)}$ denote sample estimate, "hard decision" estimate and "soft decision" estimate, respectively. $x \bmod y$ represents the modulus operation, which result is the remainder of division of $x$ by $y$; $\lfloor x \rfloor$ is the floor function (largest integer less then or equal to $x$); $\delta_{n,m}$ is the Kronecker delta function, which is 1 if $n = m$ and 0 otherwise; $E[\cdot]$ denotes expectation.

The performance results presented herein were obtained by Monte Carlo simulations using the MatLab software environment. Since the accuracy of Monte Carlo simulations strongly depend on its length (in this case, the number of transmitted blocks simulated), all simulations' length were chosen to be large enough so that the outcome was not influenced by insufficient estimated values. For example, the number of simulated blocks is always higher then 5000 and the number of bits is higher than $10^6$ ($10^7$ for the coded case).

The results presented here concern uncoded and coded bit error rate (BER) performance (similar conclusions could be drawn for the block error rate, provided that suitable channel coding schemes are employed). Perfect time synchronization between transmitted blocks associated to different users[1] and perfect channel estimation conditions are assumed in all cases. With the exception of the results shown in Chapter 5, a linear power amplification at the transmitter is also assumed. The transmitted bits are mapped into quaternary phase shift keying (QPSK) symbols under a Gray mapping rule.

Unless otherwise stated, we consider a severely time-dispersive channel characterized by the power delay profile (PDP) type C for the High Performance Local Area Network (HiperLAN/2) [11], shown in Fig. A.1 (similar results were obtained for other severely time-dispersive channels). The duration of the CIR is 1/4 times the duration of the useful part of the block (we consider block with useful part of 4 $\mu$s).

---

[1] As we will see, perfect time synchronization is not required for CP-assisted transmissions, provided that we have perfect channel estimation.

The coded BER performances are obtained by considering the use of the well-known rate-1/2 64-state convolutional code with generator polynomials given by

$$g_1(D) = 1 + D^2 + D^3 + D^5 + D^6 \qquad (1.1)$$

and

$$g_2(D) = 1 + D + D^2 + D^3 + D^6. \qquad (1.2)$$

# 2

---

# CP-Assisted CDMA Techniques

---

This chapter presents the basic principles of MC and SC modulations. This concept is then extended to SS techniques, particularly CDMA schemes such as MC-CDMA and DS-CDMA.

The chapter is organized as follows: Section 2.1 presents the basic principles of MC modulations and their relations with SC modulations. OFDM modulations and SC modulations with linear FDE and with IB-DFE are also described. Sections 2.2 and 2.3 describe the basic concepts of MC-CDMA and DS-CDMA schemes, respectively, for both the downlink and the uplink.

## 2.1 OFDM versus Single Carrier with Frequency-Domain Equalization

The basic principles of MC modulations and their relation with SC modulations are presented in this section. We also describe OFDM schemes [3] and SC-FDE schemes.

First, let us consider a linear single carrier modulation where energy associated to each symbol occupies the total transmission band. For an $N$-symbol burst the complex envelope can be written as

$$s(t) = \sum_{n=-\frac{N}{2}}^{\frac{N}{2}-1} s_n r(t - nT_s),\qquad(2.1)$$

where $r(t)$ represents the transmitted impulse, $T_s$ is the symbol duration and $s_n$ is a complex coefficient representing the $n$th symbol resulting from a direct mapping rule of the original data bits into a selected signal constellation. By applying the Fourier transform (FT) to both sides of (2.1), the following equivalent frequency-domain expression is obtained:

$$S(f) = \mathcal{F}\{s(t)\} = \sum_{n=-\frac{N}{2}}^{\frac{N}{2}-1} s_n R(f) \exp(-j2\pi f n T_s), \qquad (2.2)$$

where $R(f)$ is the FT of $r(t)$. Clearly, the transmission band associated to each symbol $s_n$ is the band occupied by $R(f)$.

Now, let us consider a multicarrier modulation, i.e., a modulation where symbols are sequentially transmitted in the frequency-domain, each one on a different subcarrier, during the time interval $T$. The spectrum of a multicarrier $N$-symbol burst can then be written as

$$S(f) = \sum_{k=-\frac{N}{2}}^{\frac{N}{2}-1} S_k R(f - kF), \qquad (2.3)$$

where $F = 1/T$ is the subcarriers' separation, $N$ the number of subcarriers and $S_k$ the $k$th frequency-domain symbol. By applying the inverse FT (IFT) to both sides of (2.3) we obtain

$$s(t) = \mathcal{F}^{-1}\{S(f)\} = \sum_{k=-\frac{N}{2}}^{\frac{N}{2}-1} S_k r(t) \exp(j2\pi t k F). \qquad (2.4)$$

The conventional frequency division multiplexing (FDM) where the spectrum associated to different symbols does not overlap is the simplest multicarrier modulation scheme. If we assume that the bandwidth associated to $R(f)$ is smaller than $F$,[1] then each symbol $S_k$ occupies a fraction $1/N$ of the total transmission band.

Comparing (2.1) with (2.3) and (2.2) with (2.4), we can conclude that multicarrier modulations can be regarded as a dual version of conventional single carrier modulations by exchanging time and frequency domains.

For single carrier modulations, the orthogonality condition between the impulses $r(t)$ associated to different symbols

$$\int_{-\infty}^{\infty} r(t - nT_s) r^*(t - n'T_s) dt = 0, \quad n \neq n' \qquad (2.5)$$

---

[1] Naturally, $F$ is the bilateral bandwidth and $F/2$ is the unilateral bandwidth.

ensures an ISI-free transmission at the receiver's matched filter output. Making use of the duality relation just referred, the orthogonality condition between the subcarriers for a multicarrier modulation can easily be written as

$$\int_{-\infty}^{\infty} R(f - kF)R^*(f - k'F)df = 0, \quad k \neq k', \tag{2.6}$$

which, from the Parseval's Theorem, is equivalent to

$$\int_{-\infty}^{\infty} |r(t)|^2 \exp(-2j\pi(k - k')Ft)dt = 0, \quad k \neq k'. \tag{2.7}$$

It is known that with linear single carrier modulations the impulses $\{r(t - iT); i = 0, 1, \ldots\}$ can be orthogonal even if they overlap in time. For example,

$$r(t) = \text{sinc}\left(\frac{t}{T_s}\right), \tag{2.8}$$

with $\text{sinc}(x) = \sin(\pi x)/(\pi x)$, satisfies the orthogonality condition (2.5). Similarly, for multicarrier modulations it is possible to verify the orthogonality between subcarriers, expressed by (2.6) (or (2.7)), even when $\{R(f - kF), k = 0, 1, \ldots, N - 1\}$ overlap in the frequency-domain. This means that we do not need to restrict ourselves to the conventional FDM case, where the spectra associated to different frequency channels do not overlap. In fact, if we have

$$R(f) = \text{sinc}\left(\frac{f}{F}\right) \tag{2.9}$$

then the corresponding time-domain impulse $r(t)$ is rectangular with duration $T = 1/F$. In this case (2.7) reduces to

$$\int_{t_0}^{t_0+T} \exp(-2j\pi(k - k')Ft)dt = 0, \quad k \neq k' \tag{2.10}$$

and, for a subcarrier separation $F$, the $N$ subcarriers are orthogonal when $T = 1/F$.

Figures 2.1 (a) and (b) show the spectrum of each OFDM individual subcarrier and the corresponding OFDM power spectral density (PSD) for $N = 16$ orthogonal subcarriers, respectively. From these figures we can observe that the orthogonal nature of the OFDM transmission results from the fact that the peak of each subcarrier's spectrum corresponds to the evenly spaced nulls of all other subcarriers' spectra with separation $F$.

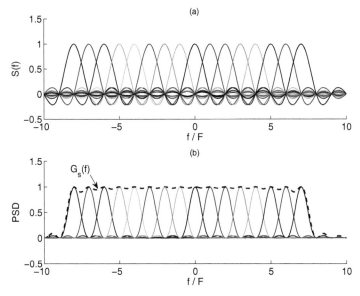

Figure 2.1 (a) Spectrum of each OFDM subcarrier; (b) OFDM PSD $(G_s(f))$ with $N = 16$ (orthogonal) subcarriers (- - -) and PSD of each subcarrier (—).

For conventional OFDM modulations the adopted impulses $r(t)$ have usually the following rectangular shape:

$$r(t) = \text{rect}\left(\frac{t}{T + T_G}\right) \tag{2.11}$$

where $T = 1/F$ and $T_G \geq 0$ represents the guard period. It will be shown in the following that this guard period can be used to cope with time-dispersive channels. Although (2.7) is not verified with impulses given by (2.11), we can say that the different subcarriers are orthogonal for the interval $[0, T]$, which will be the effective detection interval. Since

$$s^{(P)}(t) = \sum_{k=-\frac{N}{2}}^{\frac{N}{2}-1} S_k \exp\left(j2\pi \frac{kt}{T}\right) \tag{2.12}$$

is a periodic signal with period $T$, the complex envelope associated to the guard period is a repetition of the MC burst's final part, i.e.,

$$s(t) = s(t + T), \quad -T_G \leq t \leq 0. \tag{2.13}$$

Contrarily to classical FDM systems where the total frequency band is subdivided into $N$ non-overlapping frequency subchannels, each one modulated with a separate symbol and subsequently frequency multiplexed, OFDM modulations are multicarrier modulations that make a much more efficient use of bandwidth since they achieve the orthogonality conditions between different subcarriers while allowing spectral overlapping between individual subchannels.[2]

## 2.1.1 OFDM Schemes

### Transmitter Structure

The complex envelope of OFDM signals can be described as a sum of bursts with duration $T_B \geq T$, with the separation between subcarriers denoted by $F$ and the duration of the useful part equal to $T = 1/F$, i.e.,

$$s^{Tx}(t) = \sum_m s^{(m)}(t - mT_B). \tag{2.14}$$

Given (2.4), the $m$th OFDM burst can be written as

$$s^{(m)}(t) = \sum_{k=-\frac{N}{2}}^{\frac{N}{2}-1} S_k^{(m)} r(t) \exp(j2\pi k F t) = s^{(m,P)}(t) r(t), \tag{2.15}$$

where, according with (2.12),

$$s^{(m,P)}(t) = \sum_{k=-\frac{N}{2}}^{\frac{N}{2}-1} S_k^{(m)} \exp\left(j2\pi \frac{kt}{T}\right). \tag{2.16}$$

The coefficient $S_k^{(m)}$ denotes the $k$th symbol of the $m$th burst and $r(t)$, a rectangular impulse, whose duration should be greater then $1/F$ ($T_B = T + T_G \geq T = 1/F$) and appropriately adjusted so as to deal with the time-dispersive conditions of the channels as described in the following.

Assuming uncorrelated symbols, the PSD of the OFDM signal is

$$G_s(f) = \frac{1}{T_B} \sum_{k=-\frac{N}{2}}^{\frac{N}{2}-1} E\left[\left|S_k^{(m)}\right|^2\right] \left|R\left(f - \frac{k}{T}\right)\right|^2, \tag{2.17}$$

---

[2] In practice, an efficient use of the spectrum is still possible with non-overlapping FDM channels provided that $R(f)$ satisfies the 1st Nyquist criterion. However, transmitter and receiver complexities are much higher.

which converges to a rectangular impulse with bandwidth $NF$ as the number of subcarriers increases.

The complex envelope of OFDM signals given by (2.14) and (2.15) suggests a conceptual FDM transmitter structure with a bank of $N$ parallel single carrier modulators with frequencies $f_k = f_c + kF$, $k = 0, 1, \ldots, N-1$, with $f_c$ denoting the frequency of the first subcarrier. Nevertheless, for a large number of subcarriers this is not a practical structure since it would require hundreds or even thousands of local oscillators and multipliers.

However, the basic OFDM transmitter structure shown in Fig. 2.2 can be employed for OFDM schemes with a single orthogonal modulator, where the "in-phase" and "quadrature" components of each OFDM burst are obtained from a sequence of samples corresponding to the IDFT of the block to be transmitted. To understand the motivation behind this transmitter implementation consider the signal

$$s^{(P)}(t) = \sum_{k=0}^{N-1} S_k' \exp\left( j2\pi \frac{kt}{T} \right),$$ (2.18)

where the block $\{S_k'; k = 0, 1, \ldots, N-1\}$ is obtained from the original block in the frequency-domain $\{S_k; k = -N/2, -N/2+1, \ldots, N/2-1\}$ by

$$S_k' = \begin{cases} S_k, & 0 \le k \le \frac{N}{2} - 1 \\ S_{k-N}, & \frac{N}{2} \le k \le N-1, \end{cases}$$ (2.19)

as illustrated in Fig. 2.3. By applying the FT to both sides of (2.18), the expression representing the power spectrum of $s^{(P)}(t)$ is

$$S^{(P)}(f) = \mathcal{F}\{s^{(P)}(t)\} = \sum_{k=0}^{N-1} S_k' \delta\left( f - \frac{k}{T} \right).$$ (2.20)

From (2.20), it is clear that the band occupied by $s^{(P)}(t)$ is $N/T = NF$. According to the sampling theorem, $s^{(P)}(t)$ can be completely recovered from its samples taken in the interval $[0, T[$ with a sampling rate $1/T_s = N/T$, i.e.,[3]

$$s_n^{(P)} \triangleq s^{(P)}\left( \frac{nT}{N} \right) = \sum_{k=0}^{N-1} S_k' \exp\left( j2\pi \frac{kn}{N} \right) = N s_n',$$ (2.21)

---

[3] With our DFT definition, the first sample corresponds to instant (or frequency) '0' [12].

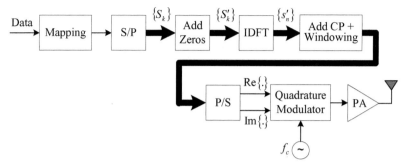

Figure 2.2 OFDM transmitter structure.

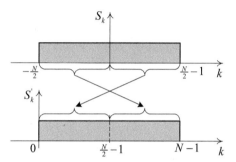

Figure 2.3 Blocks $S_k$ and $S_k'$.

$n = 0, 1, \ldots, N - 1$, where the block $\{s_n'; n = 0, 1, \ldots, N - 1\}$ is the IDFT of $\{S_k'; k = 0, 1, \ldots, N - 1\}$ with

$$s_n' = \frac{1}{N} \sum_{k=0}^{N-1} S_k' \exp\left( j2\pi \frac{kn}{N} \right). \tag{2.22}$$

Therefore, the sampled version of $s^{(P)}(t)$ in the interval $[0, T[$ corresponds, apart a scalar factor $N$, to the IDFT of the block $\{S_k'; k = 0, \ldots, N - 1\}$ in the frequency-domain, which can be efficiently implemented with the well known FFT algorithm [6].

After generating the samples of $s^{(P)}(t)$, the wave shape associated to a given burst is obtained by multiplying those samples with the samples of the "time window" $r(t)$ whose duration is higher than $T$ (see (2.15)). This means that the burst samples for an OFDM signal are given by $s_n' r_n$, with $r_n \triangleq r(nT/N)$ (note that, contrarily to the samples $s_n'$, the samples $s_n' r_n$ are not periodic).

Finally, the analog signal associated to a given OFDM burst is generated from the samples $s'_n r_n$ by digital-to-analog conversion (D/A) followed by reconstruction filtering. The complex envelope can be written as

$$s(t) = \sum_{n=-\infty}^{+\infty} s'_n r_n h_T \left( t - n \frac{T}{N} \right), \tag{2.23}$$

where $h_T(t)$ is the impulsive response of the reconstruction filter.

To simplify the reconstruction filter $h_T(t)$, usually the samples of the OFDM burst given by (2.15) are taken with a sample rate $M_{Tx} N/T > N/T$, with an oversampling factor $M_{Tx} > 1$, not necessarily integer. Usually, the original block $\{S_k; k = -N/2, -N/2+1, \ldots, N/2-1\}$ already includes $2N_I$ "idle" subcarriers (i.e., with $S_k = 0$), half of them at the beginning and the other half at the end of the burst [5]. This is in fact equivalent to oversampling an OFDM burst by a factor

$$M_{Tx} = \frac{N}{N - 2N_I} \tag{2.24}$$

with $N - 2N_I$ useful subcarriers.

When we have an oversampling factor $M_{Tx}$ for a reference burst with $N$ subcarriers, the samples of $s^{(P)}(t)$ in the interval $[0, T[$ are given by

$$s_n^{(M_{Tx})} \triangleq s^{(P)} \left( n \frac{T}{N'} \right) = \sum_{k=-\frac{N}{2}}^{\frac{N}{2}-1} S_k \exp \left( j2\pi \frac{nk}{N'} \right), \tag{2.25}$$

$n = 0, 1, \ldots, N' - 1$, where $N' = N M_{Tx}$. In this case,

$$s_n^{(M_{Tx})} = N' \left( \frac{1}{N'} \sum_{k=0}^{N-1} S'_k \exp \left( j2\pi \frac{nk}{N'} \right) \right) = N' s'_n, \tag{2.26}$$

$n = 0, 1, \ldots, N' - 1$, where the block $\{s'_n; n = 0, 1, \ldots, N' - 1\}$ is the IDFT of the block $\{S'_k; k = 0, 1, \ldots, N' - 1\}$, i.e.,

$$s'_n = \frac{1}{N'} \sum_{k=0}^{N'-1} S'_k \exp \left( j2\pi \frac{nk}{N'} \right). \tag{2.27}$$

The extended block $\{S'_k; k = 0, 1, \ldots, N' - 1\}$ is obtained by adding $N' - N$ zeros to the original block in the frequency-domain, $\{S_k; k = -N/2, -N/2+$

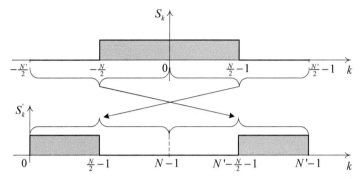

Figure 2.4 Blocks $S_k$ and $S'_k$ with $N' - N$ zeros.

$1, \ldots, N/2 - 1$}, in the following way (see Fig. 2.4):

$$S'_k = \begin{cases} S_k, & 0 \le k \le \frac{N}{2} - 1 \\ 0, & \frac{N}{2} \le k \le N' - \frac{N}{2} - 1 \\ S_{k-N'}, & N' - \frac{N}{2} \le k \le N' - 1 \end{cases} . \tag{2.28}$$

(for $M_{Tx} = 1$, (2.28) reduces to (2.19)). Again, apart a scalar factor $N'$, the sampled version of (2.18) with an oversampling factor $M_{Tx}$ corresponds to the IDFT of the block $\{S'_k; k = 0, \ldots, N' - 1\}$.

As in (2.23), the complex envelope of the analog signal associated to a given OFDM burst with an oversampling factor $M_{Tx}$ is given by

$$s^{(M_{Tx})}(t) = \sum_{n=-\infty}^{+\infty} s'_n r_n h_T \left( t - n\frac{T}{N'} \right) . \tag{2.29}$$

Although signal $s^{(M_{Tx})}(t)$ is not equal to the reference representation of the OFDM burst (given by (2.15)), the difference is very small, especially for a large number of subcarriers and/or when the oversampling factor is high, with differences manifesting mainly at the extremes vicinities of the interval occupied by $r(t)$; obviously, for $M_{Tx} \to +\infty$ the signal given by (2.29) converges to the OFDM reference burst.

For the "time window" $r(t)$, it is common to employ a square-root raised-cosine window instead of a rectangular shape window to reduce the out-of-band radiation levels on the spectrum of OFDM bursts. As shown in Fig. 2.5, the duration of the time window $r(t)$ is $T_B + T_W = T + T_G + 2T_W$, leading to an overlap of $T_W$ between adjacent bursts. This means that this raised-cosine window has a roll-off factor of $T_W/(T + T_G + T_W)$. Throughout this work

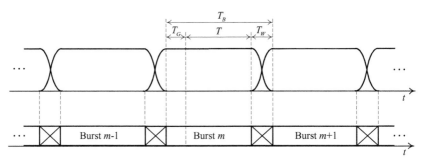

Figure 2.5  Raised-cosine time window.

we assume that $T_W = 0$, although our results are valid for other values of $T_W$ (the major impact of $T_W$ is on the PSD of the transmitted signals, something not addressed here).

### Receiver Structure

Figure 2.6 shows the OFDM receiver structure. The signal sampling is performed with a sample rate

$$\frac{1}{T_s} = \frac{N'}{T} \tag{2.30}$$

(without loss of generality, it is assumed that the reception oversampling factor is the same as the transmission oversampling factor).

Due to the multipath propagation, the received bursts will overlap. Moreover, it can be shown that the orthogonality between subcarriers can be lost. However, since the detection of OFDM signal operates on signal samples associated to a useful period of duration $T$, the use of CPs with duration, $T_G$, longer than the overall channel impulse response (which includes the impact of the transmission and detection filters as well as radio channel itself) prevents the effects of overlapping bursts in the received samples associated to the useful interval as illustrated in Fig. 2.7; this is usually referred to as the "absence of inter-symbol interference".[4] On the other hand, since the CP corresponds to a cyclic extension of each burst, this means that, the received signal associated to the useful interval $T$ is identical to the received signal when the transmitted signal corresponds, not to a sequence of OFDM bursts expressed by (2.14), but to the periodic signal $s^{(P)}(t)$ given by (2.12) (see Fig. 2.8). Therefore,

---

[4] This definition results from the fact that in OFDM literature an OFDM block is usually referred as an "OFDM symbol". The term IBI (Inter-Block Interference) would probably be more adequate.

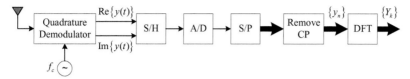

Figure 2.6 OFDM receiver structure.

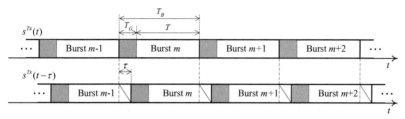

Figure 2.7 Elimination of IBI through guard periods.

the linear convolution associated to the channel is formally equivalent to a circular convolution with respect to the useful part of the OFDM block, as described in Appendix B. This means that the received sample at the $k$th subcarrier is given by

$$Y_k = H_k S_k + N_k, \tag{2.31}$$

$k = 0, 1, \ldots, N - 1$, where $H_k$ is the channel frequency response for the $k$th subcarrier and $N_k$ the Gaussian channel noise component for that subcarrier. Therefore, the channel acts like a simple multiplicative factor for each subcarrier, preserving the orthogonality between subcarriers in the useful interval; this is usually referred as the "absence of inter-channel interference" (ICI).

To avoid power and spectral degradation, the CP length should be a small fraction of the overall length of the blocks.

It should be mentioned that the equivalence between linear convolution and circular convolution associated to the use of CP-extended bursts can also be achieved by extending each burst with any known fixed sequence, including an all zeros sequence (zero padding (ZP)) [13, 14], or a pseudo noise (PN) symbol sequence, denoted PN extension or unique word (UW) which can also be used for channel estimation purposes [15]. ZP OFDM schemes can have better uncoded performances, even in the presence of deep notches in the in-band region, than CP-assisted OFDM schemes [16], and allow blind or semi-blind receivers [17, 18]. Moreover, no "useless" power is spent on the extension with ZP schemes, allowing increased power efficiency. The adopted CP or ZP length is an upper bound on the expected length of the channel

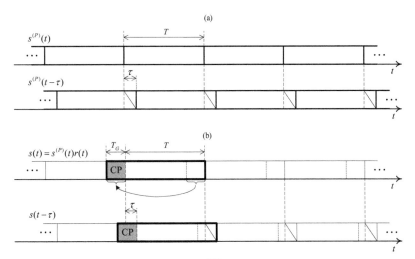

Figure 2.8 (a) Multipath channel impact on $s^{(P)}$ and (b) on the corresponding CP-extended OFDM burst.

impulse response, which can be much higher than the true channel length. In this case, ZP schemes can be a good alternative to CP-assisted schemes [13]. However, to achieve this complex receiver structures must be employed, involving the inversion and/or the multiplication of matrices whose dimensions grow with the block length [14], which are not suitable when large blocks are employed. By using overlap-add (OA) techniques, the receiver complexity is similar to the one of conventional CP-assisted schemes, but the performance is also identical [14].

In this work we will consider only CP-assisted block transmission. However, most of our results could directly be extended to ZP by considering OA schemes as well as to continuous transmission systems (which do not incorporate a CP), by using overlap-save (OS) schemes [19, 20].

Since CP-assisted transmission can be seen as a transmission over $N$ parallel non-selective subchannels (as expressed by (2.31)), channel distortion effects for an uncoded OFDM transmission can easily be compensated by the receiver depicted in Fig. 2.9. For a receiver without space diversity (i.e., $L = 1$), the equalized frequency-domain sample at the $k$th subcarrier $\tilde{S}_k$ is obtained by

$$\tilde{S}_k = F_k Y_k, \tag{2.32}$$

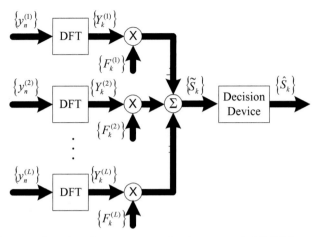

Figure 2.9 Channel distortion effects compensation for an uncoded OFDM transmission with an $L$-order space diversity receiver.

with the set of coefficients $\{F_k; k = 0, 1, \ldots, N - 1\}$ given by

$$F_k = \frac{1}{H_k} = \frac{H_k^*}{|H_k|^2}, \tag{2.33}$$

corresponding to an FDE optimized under the zero-forcing (ZF) criterion [2]. Clearly, the decision on the symbol transmitted by subcarrier $k$ can be made based on $\tilde{S}_k$.

In the case where we have $L$-order space diversity, the received sample at the $k$th subcarrier and $l$th diversity branch is given by

$$Y_k^{(l)} = S_k H_k^{(l)} + N_k^{(l)}, \tag{2.34}$$

$(l = 1, \ldots, L)$, where $H_k^{(l)}$ and $N_k^{(l)}$ denote the channel frequency response and noise term for the $k$th subcarrier and the $l$th diversity branch, respectively. The corresponding equalized sample is

$$\tilde{S}_k = \sum_{l=1}^{L} F_k^{(l)} Y_k^{(l)}, \tag{2.35}$$

where the set $\{F_k^{(l)}; k = 0, 1, \ldots, N - 1\}$ $(l = 1, \ldots, L)$ denotes the FDE coefficients associated to the $l$th diversity branch, which can be set as

$$F_k^{(l)} = \frac{H_k^{(l)*}}{\sum_{l'=1}^{L} \left| H_k^{(l')} \right|^2}. \tag{2.36}$$

Therefore, the receiver can be the one depicted in Fig. 2.9 and, from (2.34),

$$\tilde{S}_k = S_k + \frac{\sum_{l=1}^{L} H_k^{(l)*}}{\sum_{l'=1}^{L} \left| H_k^{(l')} \right|^2} N_k^{(l)}. \tag{2.37}$$

Clearly, the receiver structure of Fig. 2.9 implements a maximal-ratio combining (MRC) [21] diversity scheme for each subcarrier $k$.

## 2.1.2 Single Carrier with Frequency-Domain Equalization

One of the major problems of OFDM schemes relatively to SC schemes is the strong envelope fluctuations and PMEPR of the transmitted signals making them vulnerable to transmitter nonlinearities, namely those associated to the power amplifier. For this reason it is desirable to employ SC modulations, especially at the uplink transmission, allowing cheaper and more efficient high-power amplifiers to be used in user terminals.

When conventional SC modulations are employed on digital communications systems requiring transmission rates of several Mbits/s over severely time-dispersive channels the transmission bandwidth becomes much higher than the coherence bandwidth of the radio channel, leading to severe signal distortion levels unless high complexity receivers are employed. The optimum receiver structure corresponds to the well-known Viterbi equalizer [1]. However, since complexity grows exponentially with the channel impulse response length, a Viterbi equalizer is not practical for moderately and severely time-dispersive channels where the channel impulse response expands over several tens or even several hundreds of symbols.

Typically, time-domain equalizers comprehending one or more transversal filters are used to mitigate the hard ISI problem, leading to receivers much simpler than Viterbi equalizers at the expense of some performance degradation. When nonlinear structures such as DFEs [2] are employed to compensate the distortion caused by channel frequency selectivity, their BER performance is usually much better than that of a linear equalizer and can come close to that of an optimum sequence detector implemented by the Viterbi algorithm. However, signal processing complexity of time-domain equalizers (measured in terms of the number of arithmetic operations per data symbol) increases at least linearly with the number of data symbols spanned by the CIR. This means very high receiver complexity (and the inherent power consumption) with exorbitant signal processing requirements, and, consequently, making time-domain equalizers not suitable for severely time-dispersive channels.

Following the success of OFDM applications in wireless systems, it was shown that SC modulations could benefit from FDE techniques if a blockwise transmission approach is used, as with current MC-based OFDM implementation [22–25].

To understand the SC-FDE concept, let us consider an SC-based block transmission with $N$ useful modulation symbols per block $\{s_n; n = 0, 1, \ldots, N - 1\}$ resulting from a direct mapping of the original data into a selected signal constellation, plus a suitable CP. The receiver can have the structure represented in Fig. 2.10, where it is assumed that after being down-converted and filtered, the signal is sampled and A/D converted. The resulting signal is S/P converted and the CP samples are removed. For a receiver with no space diversity ($L = 1$), this leads to the received time-domain samples $\{y_n; n = 0, 1, \ldots, N-1\}$. These samples are passed to the frequency-domain by an $N$-point DFT, leading to the corresponding frequency-domain samples $\{Y_k; k = 0, 1, \ldots, N - 1\}$, where

$$Y_k = H_k S_k + N_k. \tag{2.38}$$

For an FDE optimized under the ZF criterion the equalized frequency-domain samples, given by (2.32), can be obtained with the set of coefficients $F_k$ given by (2.33). However, for a typical frequency-selective channel, deep notches in the channel frequency response lead to significant noise enhancement effects when ZF criterion is employed. To minimize the combined effect of ISI and channel noise, the equalized samples $\{\tilde{S}_k; k = 0, 1, \ldots, N - 1\}$ are usually obtained with the coefficients $\{F_k; k = 0, 1, \ldots, N - 1\}$ optimized under the minimum mean square error (MMSE) criterion [2], given by

$$F_k = \frac{H_k^*}{\beta + |H_k|^2}, \tag{2.39}$$

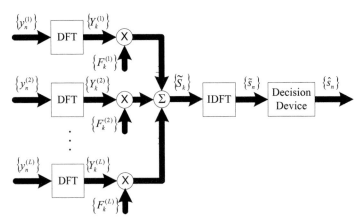

Figure 2.10  SC-FDE receiver structure with an $L$-order space diversity receiver.

$k = 0, 1, \ldots, N-1$, where $\beta$ is the inverse of the signal-to-noise ratio (SNR), given by

$$\beta = \frac{\sigma_N^2}{\sigma_S^2},\qquad(2.40)$$

with

$$\sigma_N^2 = \frac{E\left[|N_k|^2\right]}{2}\qquad(2.41)$$

and

$$\sigma_S^2 = \frac{E\left[|S_k|^2\right]}{2}\qquad(2.42)$$

denoting the variances of the real and imaginary parts of the channel noise components $\{N_k; k = 0, 1, \ldots, N - 1\}$ and the data samples $\{S_k; k = 0, 1, \ldots, N - 1\}$, respectively.

Since, for SC modulations the data contents of a given block are transmitted in the time-domain, the equalized samples $\{\tilde{S}_k; k = 0, \ldots, N - 1\}$ are converted back to the time-domain by an IDFT operation leading to the block of time-domain equalized samples $\{\tilde{s}_n; n = 0, \ldots, N - 1\}$. These equalized samples will then be used to make decisions on the transmitted symbols.

In the case of an $L$-branch diversity scenario as depicted in Fig. 2.10, the frequency-domain samples at the FDE's output are given by (2.35), where the set $\{F_k^{(l)}; k = 0, 1, \ldots, N - 1\}$ ($l = 1, .., L$) can be set as (2.36) under the ZF criterion. Under "equal noise level" conditions (i.e., $\sigma_n^{(1)} = \ldots = \sigma_n^{(L)} \triangleq \sigma_n$, with $(\sigma_n^{(l)})^2$ denoting the variance of the input noise samples at branch *l*) and an MMSE criterion, it can be easily shown that the optimized FDE coefficients are given by [26]

$$F_k^{(l)} = \frac{H_k^{(l)*}}{\beta + \sum_{l'=1}^{L} \left| H_k^{(l')} \right|^2},$$ (2.43)

$l = 1, 2, \ldots, L.$

Contrarily to the OFDM schemes where ZF and MMSE criteria yield the same BER performance in the uncoded case [26], for the SC-FDE these criteria are only equivalent when the channel frequency response across the transmission bandwidth is constant (i.e., when $H_0^{(l)} = \ldots = H_{N-1}^{(l)} \triangleq H^{(l)}, l = 1, \ldots, L$). An FDE optimized under the MMSE criteria does not attempt to fully invert the channel when we have a deep fade, reducing noise enhancement effects and allowing better performances.

Figure 2.11 shows the compensation of channel distortion effect when the FDE coefficients $\{F_k; k = 0, \ldots, N - 1\}$ are optimized under the ZF or MMSE criteria (given by (2.33) and (2.39), respectively) for a frequency-selective channel realization with deep in-band notches. Clearly, with ZF criterion the channel is completely inverted, resulting in a perfect equalized channel after the FDE, while MMSE criterion leads to an imperfect channel equalization. However, as shown in Fig. 2.12, in spite of perfectly equalizing the channel, ZF criterion significantly enhances the channel noise at subchannels with local deep notches, while with the MMSE criterion the noise-dependent term $\beta$ in (2.39) avoids noise enhancement effects for very low values of the local channel frequency response. This noise enhancement effect can also be observed from Figs. 2.13 and 2.14 where it is shown the equalized samples $\{\tilde{s}_n; n = 0, 1, \ldots, N - 1\}$ at the FDE output under the ZF or the MMSE criteria, with and without the presence of the channel noise, respectively. From these figures we can observe that the perfect channel inversion obtained with the ZF criterion leads to the exact values of the data samples when no channel noise is present, contrarily to the MMSE criterion at the same conditions. However, in the presence of the channel noise, the noise enhancement effect with the ZF causes higher spread of the equalized samples around the exact values of the data samples, while with MMSE criterion those samples remain closer to the true data samples values.

Figure 2.15 shows the potential benefits of space diversity (with $L = 2$) when the FDE coefficients, $\{F_k^{(l)}; k = 0, \ldots, N - 1\}$ ($l = 1, 2$), are again optimized under the ZF and MMSE criteria (given by (2.36) and (2.43),

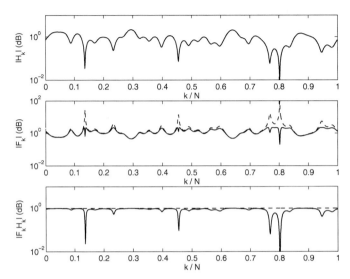

Figure 2.11  Channel frequency response (top), FDE coefficients optimized under the ZF (- - -) and MMSE criteria (—) (middle) and corresponding channel distortion effects compensation (bottom).

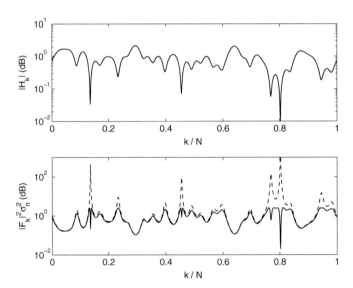

Figure 2.12  Channel frequency response (top) and noise power at the output of an FDE optimized under the ZF (- - -) and MMSE (—) criteria (bottom).

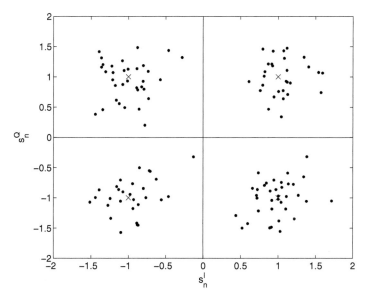

Figure 2.13 Equalized samples $\{\tilde{s}_n; n = 0, 1, \ldots, N - 1\}$ with an FDE optimized under the ZF criterion: ($\bullet$) with noise, ($\times$) without noise.

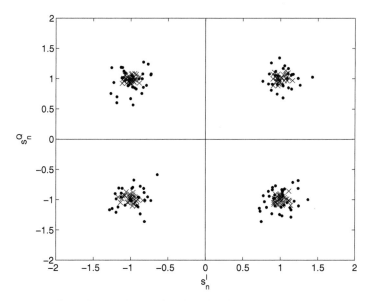

Figure 2.14 As in Fig. 2.13 but with the MMSE criterion.

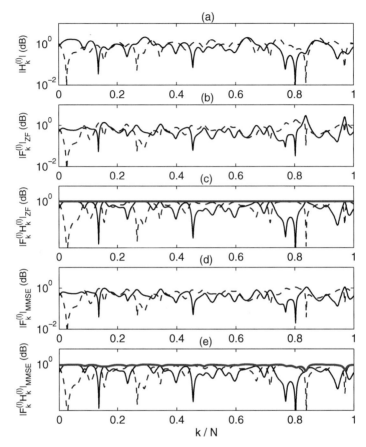

Figure 2.15  (a) Channel frequency responses for each branch of a two-branch space diversity FDE; (b) FDE coefficients optimized under the ZF criterion associated to each diversity branch; (c) ZF channel distortion effects compensation; (d) FDE coefficients optimized under the MMSE criterion associated to each diversity branch; (e) MMSE channel distortion effects compensation (tick line in (c) and (e) corresponds to $\sum_l |F_k^{(l)} H_k^{(l)}|$).

respectively). From this figure, it is clear that, since there is a very low probability of having deep notches at both receiver branches simultaneously for any $k$ (assuming uncorrelated receive antennas), the resulting combination of the FDE outputs from each diversity branch allows must better compensation of the channel effects, practically eliminating the high channel attenuation for some subcarriers in each branch. In fact, as we increase the diversity order we reduce the equalization requirements since the overall channel is closer to a frequency-flat channel.

For comparison purposes, Fig. 2.16 shows the block diagrams for the transmission chains of OFDM and SC-FDE options. From this figure, it is clear that both schemes are closely related, and to a certain extent compatible [27]. Their overall signal processing effort, measured in terms of DFT/IDFT blocks, are similar, with the only difference being the shift of the IDFT block from the OFDM transmitter to the SC-FDE receiver. In fact, not just the complexity of the SC-FDE is similar to that of OFDM, as it leads to better uncoded performance under similar equalization complexity levels and average power requirements, even when space diversity receivers are employed [22]. For these reasons, SC modulations option have become an interesting alternative to the usually recommended OFDM-type schemes for broadband wireless systems.

However, although from Fig. 2.16 both modulations choices may seem nearly equivalent, there are pros and cons in both sides that, for the sake of a fair comparison, must not be neglected [22]. One of the most important issue that must be taken into consideration is the already mentioned strong envelope fluctuations and high PMEPR of MC signals with a large number of subcarriers, leading to power amplification difficulties [22, 23]. This requires the use of highly linear power amplifiers at the transmitter and/or with increased power backoff than with comparable SC signals. This aspect is particularly important for the uplink transmission since low-cost and low-consumption power amplifiers are desirable at the MT. However, even when suitable signal processing techniques are employed to reduce the envelope fluctuations of MC signals [28–32], the resulting envelope fluctuations are still higher than with the corresponding SC schemes.

Nevertheless, when the wireless network includes fixed terminal (e.g. within BS and/or for broadcasting systems) MC schemes are excellent candidates. Having in mind the compatibility between SC-FDE and MC options we can choose an SC-FDE scheme, exhibiting low envelope fluctuations, for the uplink and an MC scheme for the downlink. This means an implementation advantage for the MTs, where simple SC transmissions and MC reception functions are then carried out. The implementation charge is concentrated in the BSs (where increased power consumption and cost are not so critical), concerning both the signal processing effort and the power amplification difficulties [22, 23].

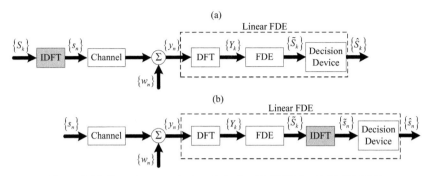

Figure 2.16 Generic transmission chain for (a) OFDM and (b) SC-FDE.

## 2.1.3 IB-DFE Receivers

It is well known that DFEs can significantly outperform linear equalizers (in fact, DFEs include as special case linear equalizers). As it was already mentioned, time-domain DFE [2] have good performance/complexity tradeoffs, provided that the channel impulse response is not too long. However, if the channel impulse response expands over a large number of symbols (such as in the case of severely time-dispersive channels), conventional time-domain DFEs are too complex. For this reason, an hybrid time-frequency SC-DFE was proposed in [33], employing a frequency-domain feedforward filter and a time-domain feedback filter. This hybrid time-frequency-domain DFE has a better performance than a linear FDE. However, as with conventional time-domain DFEs, it can suffer from error propagation, especially when the feedback filters have a large number of taps. A promising IB-DFE approach for SC transmission was proposed in [10] and extended to transmit/diversity scenarios in [34, 35]. Within these IB-DFE schemes, both the feedforward and the feedback parts are implemented in the frequency-domain, as depicted in Fig. 2.17.

Let us consider an $L$-order space diversity IB-DFE. For a given iteration $i$, the output samples are given by

$$\tilde{S}_k^{(i)} = \sum_{l=1}^{L} F_k^{(l,i)} Y_k^{(l)} - B_k^{(i)} \hat{S}_k^{(i-1)}, \qquad (2.44)$$

where $\{F_k^{(l,i)}; k = 0, 1, \ldots, N-1\}$ $(l = 1, \ldots, L)$ and $\{B_k^{(i)}; k = 0, 1, \ldots, N-1\}$ denote the feedforward and feedback equalizer coefficients, respectively, and $\{\hat{S}_k^{(i-1)}; k = 0, 1, \ldots, N-1\}$ is the DFT of the hard-decision block

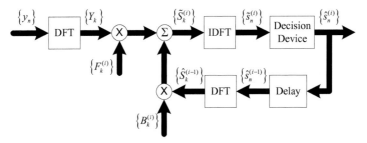

Figure 2.17  IB-DFE receiver structure.

$\{\hat{s}_n^{(i-1)}; n = 0, 1, \ldots, N - 1\}$, of the $(i - 1)$th iteration, associated to the transmitted time-domain block $\{s_n; n = 0, 1, \ldots, N - 1\}$.

The forward and backward IB-DFE coefficients $\{F_k^{(l,i)}; k = 0, 1, \ldots, N - 1\}$ $(l = 1, \ldots, L)$ and $\{B_k^{(i)}; k = 0, 1, \ldots, N - 1\}$, respectively, are chosen so as to maximize the "signal-to-interference plus noise ratio" (SINR). In [35] it is shown that the optimum feedforward and feedback coefficients are given by

$$F_k^{(l,i)} = \frac{H_k^{(l)*}}{\beta + \left(1 - \left(\rho^{(i-1)}\right)^2\right) \sum_{l'=1}^{L} \left|H_k^{(l')}\right|^2}, \qquad (2.45)$$

$l = 1, 2, \ldots, L$, and

$$B_k^{(i)} = \rho^{(i-1)} \left(\sum_{l'=1}^{L} F_k^{(l',i)} H_k^{(l')} - 1\right), \qquad (2.46)$$

respectively, where $\beta$ is given by (2.40) and the correlation coefficient

$$\rho^{(i)} = \frac{E\left[s_n^* \hat{s}_n^{(i)}\right]}{E\left[|s_n|^2\right]} \qquad (2.47)$$

is, as it will be seen later, a key parameter for the good performance of the receiver. Since the feedback loop takes into account not just the hard-decisions for each block, but also the overall block reliability, the error propagation problem is significantly reduced. Consequently, the IB-DFE techniques offer much better performances than the non-iterative methods [10,34]. In fact, the IB-DFE schemes can be regarded as low complexity turbo equalizers [36] since the feedback loop uses the equalizer outputs instead of the channel decoder outputs. For the first iteration we do not have any information about

$s_n$, which means that $\rho = 0$, $B_k = 0$ and $F_k$ are given by (2.43). Therefore the IB-DFE reduces to a linear FDE.

## 2.2 MC-CDMA Systems with Linear FDE

As we saw in the previous section, block transmission techniques with appropriate cyclic extension appended to each block and employing FDE techniques are suitable for high data rate transmissions over severely time-dispersive channels.

SS techniques, particularly coded division multiple access (CDMA) schemes have been considered to be an appropriate multiple access scheme, since they provide higher capacity and flexibility than conventional access techniques such as time-division multiple access (TDMA) and frequency-division multiple access (FDMA). With CDMA schemes a number of users simultaneously access a channel by modulating and spreading their information-bearing signals with preassigned signature sequences (spreading codes), resulting in signal overlapping, in both the time and in the frequency domains. Moreover, since users transmit continuously, the peak power requirements for the amplifiers are significantly reduced. For these reasons, CDMA schemes have been adopted in many communication and navigation systems over the past decades.

Multicarrier coded division multiple access (MC-CDMA) schemes [9, 37] combine a CDMA scheme with an OFDM modulation so as to allow high transmission rates over severely time-dispersive channels without the need of complex receiver implementations. Although there are several ways of combining CDMA with MC modulations, sometimes with misleading designations, the most frequent types of combination found in literature are designated as MC-CDMA [38–40], MC-DS-CDMA (Multicarrier Direct-Sequence CDMA) [41] and MT-CDMA (Multitone CDMA) [42]. MC-CDMA is used to describe the case where the transmitter spreads the original information-bearing signals over different subcarriers using frequency-domain spreading codes. This means that a fraction of each data symbol corresponding to a chip of the spreading code is transmitted through a different subcarrier. In MC-DS-CDMA, the transmitter spreads the data stream using a given spreading code in the time-domain so that the resulting spectrum of each subcarrier can satisfy the orthogonality condition with the minimum subcarrier separation. In MT-CDMA the data stream is also spread in the time-domain but in this case the orthogonality condition of each subcarrier is verified before the spreading operation. Therefore, the resulting spectrum of each subcarrier no longer satisfies the orthogonality condition.

In the following we will only consider the MC-CDMA variant. As with conventional OFDM, an appropriate CP is appended to each transmitted block. Since the spreading is made in the frequency-domain, the time synchronization requirements are much lower than with conventional direct sequence CDMA schemes (DS-CDMA). Moreover, by spreading a given data symbol over several subcarriers there is an inherent diversity effect in frequency-selective channels. Since the transmission over time-dispersive channels destroys the orthogonality between spreading codes, some kind of FDE is required at the receiver before the despreading operation [43].

## 2.2.1 Downlink Transmission

This subsection considers the use of MC-CDMA schemes in the downlink transmission. Let us start with a single user transmission employing CP-assisted block transmission techniques as depicted in Fig. 2.18(a). Within that structure, a sequence of $M$ data symbols selected from a given constellation with an appropriate mapping rule is serial-to-parallel converted. The corresponding block of data symbols $\{A_m; m = 0, 1, \ldots, M - 1\}$ is submitted to a frequency-domain spreading operation by multiplication with a spreading sequence $\{C_k; k = 0, 1, \ldots, N - 1\}$, where $N = KM$, with $K$ denoting the spreading factor (also known as spreading gain or processing gain). It is assumed that $C_k$ belongs to a QPSK constellation[5] and, without loss of generality, that $|C_k| = 1$. To improve diversity gains, different chips associated to a given data symbol should not be transmitted by adjacent subcarriers. Therefore, the frequency-domain block of chips to be transmitted, $\{S_k; k = 0, 1, \ldots, N - 1\}$, is an interleaved version of the block $\{S'_k; k = 0, 1, \ldots, N - 1\}$, given by

$$S'_k = C_k A_{\lfloor k/K \rfloor} \tag{2.48}$$

($\lfloor x \rfloor$ denotes "larger integer not higher than $x$"). This means that chips associated to a given data symbol are uniformly spread within the transmission band (corresponding to employ a rectangular $K \times M$ interleaver, i.e., the chips are spaced by $M$ subcarriers), as illustrated in Fig. 2.19. The block

---

[5] In fact, the spreading sequence could be selected from any PSK constellation without changing the results.

(a)

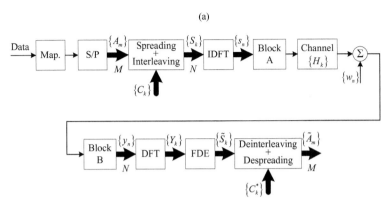

(b)

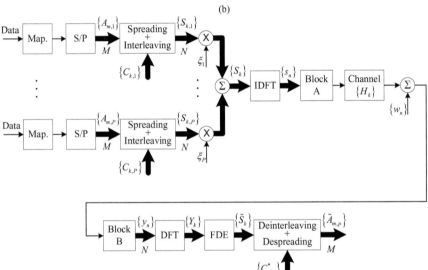

'Block A' : Add CP, Windowing, P/S conversion and 'Tx'.
'Block B' : 'Rx', S/P conversion, CP removal.

Figure 2.18  MC-CDMA downlink transmission models: (a) single user and (b) multiuser.

$\{S_k; k = 0, 1, \ldots, N - 1\}$ can be writhen as

$$S_k = C_k A_{k \bmod M}, \tag{2.49}$$

$k = 0, 1, \ldots, N - 1$.

As with other CP-assisted techniques, the CP is removed at the receiver and the resulting time-domain block $\{y_n; n = 0, 1, \ldots, N-1\}$ is passed to the frequency-domain, leading to the block $\{Y_k; k = 0, 1, \ldots, N - 1\}$, given by

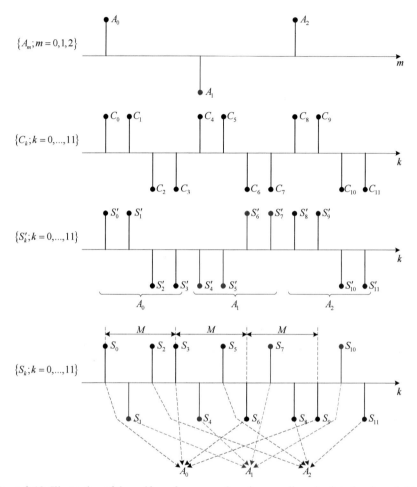

Figure 2.19 Illustration of the uniform frequency-domain spreading plus interleaving of chips associated to a given data symbol for $M = 3$ symbols and a spreading factor of $K = 4$.

(2.38). Since we have $K$ frequency replicas for each transmited data symbol, the optimum receiver corresponds to an MRC [9] diversity scheme, on each subcarrier $k$, and the data symbols $\{A_m; m = 0, 1, \dots, M - 1\}$ can then be estimated from the equalized samples $\{\tilde{A}_m; m = 0, 1, \dots, M - 1\}$, with

$$\tilde{A}_m = \sum_{k' \in \Psi_m} Y_{k'} H_{k'}^* C_{k'}^*, \qquad (2.50)$$

with $\Psi_m = \{m, m + M, \ldots, m + (K - 1)M\}$ denoting the set of frequencies employed to transmit the $K$ chips associated to the $m$th data symbol. Formally, this is equivalent to have

$$\tilde{A}_m = \sum_{k' \in \Psi_m} \tilde{S}_{k'} C_{k'}^*, \tag{2.51}$$

with

$$\tilde{S}_k = F_k Y_k, \tag{2.52}$$

i.e., our receiver can be regarded as an FDE with $F_k = H_k^*$, followed by a despreading procedure.

Let us consider now a transmission involving $P$ users as depicted in Fig. 2.18(b). For the sake of simplicity it is assumed that all users have the same spreading factor and the same rate. In this case, the frequency-domain chips to be transmitted are given by

$$S_k = \sum_{p=1}^{P} \xi_p S_{k,p} \tag{2.53}$$

($k = 0, 1, \ldots, N - 1$), where $\xi_p$ is an appropriate weighting coefficient for power control purposes (the power associated to the $p$th spreading code (or user) is proportional to $\xi_p^2$). As with single user transmission, the block $\{S_{k,p}; k = 0, 1, \ldots, N - 1\}$ is an interleaved version of the block $\{S'_{k,p}; k = 0, 1, \ldots, N - 1\}$ (as illustrated in Fig. 2.19, but with $S'_k$ and $S_k$ replaced by $S'_{k,p}$ and $S_{k,p}$, respectively), where

$$S'_{k,p} = C_{k,p} A_{\lfloor k/K \rfloor, p} \tag{2.54}$$

is the $k$th chip for the $p$th spreading code, $\{A_{m,p}; m = 0, 1, \ldots, M - 1\}$ denotes the block of data symbols associated to the $p$th spreading code and $\{C_{k,p}; k = 0, 1, \ldots, N - 1\}$ is the corresponding spreading sequence. An orthogonal spreading with $K$-length Walsh–Hadamard sequences [7] is assumed, with $C_{k,p}$ belonging to a QPSK constellation and $|C_{k,p}| = 1$. The $k$th chip for the $p$th spreading code to be transmitted is given by

$$S_{k,p} = C_{k,p} A_{k \bmod M, p}. \tag{2.55}$$

Since we are considering the downlink transmission, all spreading codes face the same channel. The block of data symbols $\{A_{m,p}; m = 0, 1, \ldots, M - 1\}$

can then be estimated by deinterleaving and despreading the block $\{\tilde{S}_k; k = 0, 1, \ldots, N - 1\}$, with $\tilde{S}_k$ given by (2.52), i.e., from

$$\tilde{A}_{m,p} = \sum_{k' \in \Psi_m} \tilde{S}_{k'} C^*_{k',p}. \tag{2.56}$$

Following the same reasoning as for the single user transmission, the natural solution for the FDE coefficients $\{F_k; k = 0, 1, \ldots, N - 1\}$ would be to explore the inherent frequency-domain diversity effect resulting from the combination of MC modulation and CDMA. In fact, since the same data symbol is transmitted by $K$ frequency replicas, one could argue that these $K$ replicas should be combined in order to obtain a better estimate of the transmitted symbol using detection techniques inspired from space diversity combining such as MRC or EGC [9, 21, 44]. However, since detection of MC-CDMA signals can be regarded as a pure channel equalization problem, detection based on MRC and EGC not only does not help equalize the channel, as it also may increase multiuser interference (MUI) and, as a consequence, significantly degrade the BER performance [43]. For instance, Figs. 2.20 and 2.21 show the average BER performance for an MC-CDMA transmission with Walsh–Hadamard spreading sequences and $N = K = 256$ subcarriers over the strong frequency-selective channel described in Section 1.3, using an MRC and an EGC-based equalizers, respectively. The BERs are expressed as a function of $E_b/N_0$, where $E_b$ denotes the average bit energy and $N_0$ the one-sided PSD of the channel noise. Data symbols are selected from a QPSK modulation under a Gray mapping rule. For comparison proposes, we also include the corresponding single user (SU) performance, defined as

$$P_{b,SU} = \frac{1}{M} \sum_{m=0}^{M-1} E \left[ Q \left( \sqrt{\frac{2E_b}{N_0} \frac{1}{K} \sum_{k \in \Psi_m} \sum_{l=1}^{L} \left| H_k^{(l)} \right|^2} \right) \right], \tag{2.57}$$

where the expectation is over the set of channel realizations (it is assumed that $E[|H_k^{(l)}|^2] = 1$, for any $k$), and AWGN channel performance

$$P_b = Q \left( \sqrt{\frac{2LE_b}{N_0}} \right). \tag{2.58}$$

From Fig. 2.20 we can observe that, while the MRC equalizer is optimum for a single user scenario, it leads to significant performance degradation as the

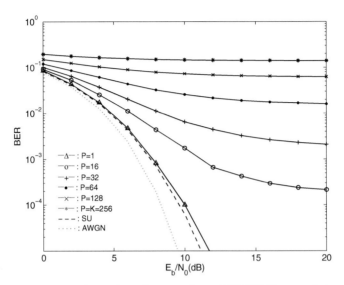

Figure 2.20 Average uncoded BER performance for an MC-CDMA transmission with $N = K = 256$ and $P = 1, 16, 32, 64, 128$ and 256 users for an MRC equalizer.

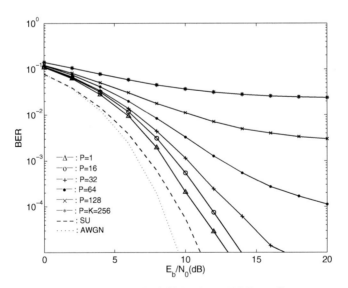

Figure 2.21 As in Fig. 2.20 but for an EGC equalizer.

number of users increase due to MUI. We can also observe from Fig. 2.21 that the EGC equalizer leads to better performance relatively to the MRC equalizer, although the performance is still very poor.

To understand this, let us consider that the equalizer coefficients $\{F_k; k = 0, 1, \ldots, N - 1\}$, are set according to an EGC scheme, i.e.,

$$F_k = \frac{H_k^*}{|H_k|}, \tag{2.59}$$

$k = 0, \ldots, N - 1$. Now, if we consider the combined channel frequency response after the FDE, i.e.,

$$H_k F_k = |H_k|, \tag{2.60}$$

we can conclude that, since the combined frequency response is real-valued, this type of combination technique simply performs channel phase equalization, leaving channel frequency-selectivity unchanged. This means that orthogonality between spreading codes cannot be recovered and serious performance degradation can be expected due to MUI.

Let us see the MRC scheme. In this case,

$$F_k = H_k^*, \tag{2.61}$$

$k = 0, \ldots, N - 1$, and the combined frequency response after the FDE is

$$H_k F_k = |H_k|^2. \tag{2.62}$$

From (2.62), it becomes clear that MRC not only does not counteract the frequ-ency-selectivity of the channel, as indeed, enhances the depth of the notches in the channel frequency response which significantly deteriorates the performance with respect to EGC scheme.

Therefore, since detection of these MC-CDMA signals could be regarded as a channel equalization problem, the FDE coefficients can be optimized under the MMSE criterion, given by (2.39), to prevent noise enhancement effects associated to a ZF optimization as with SC-FDE.

To illustrate the relative BER performance of the different MC-CDMA detection techniques, Fig. 2.22 shows some simulation results of an MC-CDMA transmission in the same conditions as in Figs. 2.20 and 2.21. The number of users is $P = K = 256$ (i.e., a fully loaded system). From this figure it is clear that MRC receiver is unusable in these conditions since it leads to significant BER degradation. The adoption of an EGC receiver

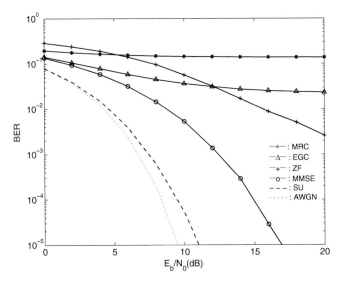

Figure 2.22 Average uncoded BER performance for an MC-CDMA transmission with $N = K = P = 256$ for different types of equalizers (MRC, EGC, ZF and MMSE).

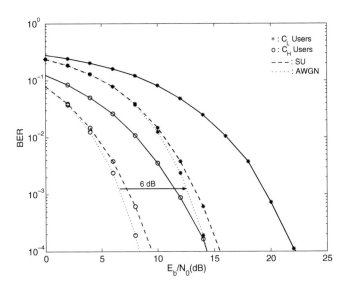

Figure 2.23 Average uncoded BER performance for the MMSE-FDE for $C_L$ and $C_H$ users as a function of the $E_b/N_0$ of $C_H$ users (average power of $C_L$ users 6 dB below the average power of $C_H$ users).

improves the performance relatively to MRC receiver but the degradation remains very large with respect to the corresponding SU performance and the AWGN channel performance. Finally, ZF and MMSE equalizers lead to better performance, with substantially advantage to the MMSE equalizer, confirming that signal detection in MC-CDMA can be regarded as a pure channel equalization problem. However, an FDE optimized under the MMSE criterion does not perform an ideal channel inversion. Therefore, when this type of equalizer is employed we are not able to fully orthogonalize the different spreading codes. This means severe interference levels leading to a performance that is still far from the SU performance, especially for fully loaded systems and/or when different powers are assigned to different spreading codes. To illustrate this, Fig. 2.23 shows the performance results for the MMSE-FDE for the same scenario of Fig. 2.22 (i.e., a fully loaded system), but when different powers are assigned to different users. We consider two classes of users denoted $C_L$ (low-power users) and $C_H$ (high-power users), with $K/2 = 128$ users in each class, where the average power of $C_H$ users is 6 dB above the average power of the $C_L$ users. From this figure it is clear that $C_L$ users face strong interference levels.

## 2.2.2 Uplink Transmission

The uplink transmission in MC-CDMA systems is more challenging due to the fact that the signals associated to different users are affected by different propagation channels. Since the orthogonality between spreading codes associated to different users is lost, we can have severe interference levels which lead to significant performance degradation, especially for fully loaded systems and/or when different powers are assigned to different spreading codes.

Let us consider an uplink transmission in an MC-CDMA system employing frequency-domain spreading involving $P$ users (MTs), transmitting independent data blocks, as depicted in Fig. 2.24. After serial-to-parallel conversion of the $M$ data symbols associated to a given user, the corresponding block $\{A_{m,p}; m = 0, 1, \ldots, M - 1\}$ is submitted to a frequency-domain spreading and interleaving operations resulting in the block of chips $\{S_{k,p}; k = 0, 1, \ldots, N - 1\}$ to be transmitted by the $p$th user, with $N = KM$ and

$$S_{k,p} = C_{k,p} A_{k \bmod M, p} \tag{2.63}$$

denoting the $k$th frequency-domain chip associated to the $p$th user. Since we are considering an uplink (asynchronous) transmission, a pseudo-random

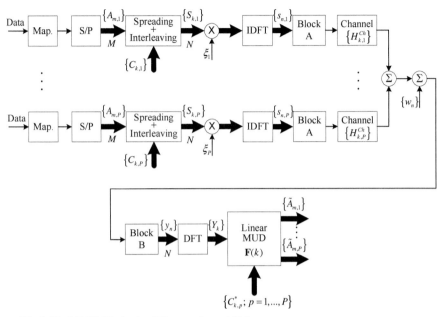

'Block A' : Add CP, Windowing, P/S conversion and 'Tx'.
'Block B' : 'Rx', S/P conversion, CP removal.

Figure 2.24  MC-CDMA uplink transmission model.

spreading is assumed (again with $C_{k,p}$ belonging to a QPSK constellation and $|C_{k,p}| = 1$) since this type of sequences present better cross-correlation properties when compared with orthogonal sequences used in an asynchronous environment. The signal received at the BS is sampled at the chip rate (the generalization for multiple samples per chip is straightforward) and the CP is removed, leading to the time-domain block $\{y_n; n = 0, 1, \ldots, N - 1\}$. After discarding the samples associated to the CP, $\{y_n; n = 0, 1, \ldots, N - 1\}$ is passed to the frequency-domain, leading to the block $\{Y_k; k = 0, 1, \ldots, N - 1\}$ with

$$Y_k = \sum_{p=1}^{P} \xi_p \, S_{k,p} \, H_{k,p}^{Ch} + N_k = \sum_{p=1}^{P} \xi_p \, C_{k,p} \, A_{k \bmod M, p} \, H_{k,p}^{Ch} + N_k$$

$$= \sum_{p=1}^{P} A_{k \bmod M, p} \, H_{k,p} + N_k, \tag{2.64}$$

where $\xi_p$ is an appropriate weighting coefficient that accounts for the propagation losses of the $p$th user, $H_{k,p}^{Ch}$ denote the channel frequency response for the $p$th user, at the $k$th subcarrier, $N_k$ the corresponding channel noise and

$$H_{k,p} = \xi_p H_{k,p}^{Ch} C_{k,p} \tag{2.65}$$

can be regarded as the "overall" channel frequency response for the $p$th user, at the $k$th subcarrier.

By defining the length-$K$ column vector with the received frequency-domain samples

$$\mathbf{Y}(k) = \begin{bmatrix} Y_k \\ Y_{k+M} \\ \vdots \\ Y_{k+(K-1)M} \end{bmatrix}, \tag{2.66}$$

and the length-$P$ column vector

$$\mathbf{A}(k) = \begin{bmatrix} A_{k \bmod M,1} \\ \vdots \\ A_{k \bmod M,P} \end{bmatrix}, \tag{2.67}$$

with the $m$th data symbol of each user, (2.64) can be written in matrix form as

$$\mathbf{Y}(k) = \mathbf{H}^T(k)\mathbf{A}(k) + \mathbf{N}(k), \tag{2.68}$$

where

$$\mathbf{N}(k) = \begin{bmatrix} N_k \\ N_{k+M} \\ \vdots \\ N_{k+(K-1)M} \end{bmatrix} \tag{2.69}$$

denotes the length-$K$ column vector with the noise samples associated to the set of frequencies $\Psi_m$ employed in the transmission of the $m$th data symbol of each user. In (2.68), $\mathbf{H}(k)$ is the $P \times K$ "overall" channel frequency response matrix associated to $\mathbf{A}(k)$, i.e.,

$$\mathbf{H}(k) = \begin{bmatrix} \mathbf{H}_1(k) \\ \vdots \\ \mathbf{H}_P(k) \end{bmatrix} = \begin{bmatrix} H_{k,1} & \cdots & H_{k+(K-1)M,1} \\ \vdots & & \vdots \\ H_{k,P} & \cdots & H_{k+(K-1)M,P} \end{bmatrix}, \tag{2.70}$$

with lines associated to the different users and columns associated to the set of frequencies $\Psi_m$, where

$$\mathbf{H}_p(k) = \begin{bmatrix} H_{k,p} & H_{k+M,p} & \cdots & H_{k+(K-1)M,p} \end{bmatrix}. \qquad (2.71)$$

To detect the $m$th data symbol of the $p$th user we will use the set of subcarriers $\Psi_m = \{m, m + M, \ldots, m + (K-1)M\}$. For a linear FDE, the equalized samples vector $\tilde{\mathbf{A}}(k)$, associated to $\mathbf{A}(k)$, is given by

$$\tilde{\mathbf{A}}(k) = \mathbf{F}^T(k)\mathbf{Y}(k), \qquad (2.72)$$

where $\tilde{\mathbf{A}}(k)$ is defined in a similar form as $\mathbf{A}(k)$ by (2.67) and $\mathbf{F}(k)$ is a $K \times P$ matrix with the FDE coefficients given by

$$\mathbf{F}(k) = \begin{bmatrix} \mathbf{F}_1(k) & \cdots & \mathbf{F}_P(k) \end{bmatrix} = \begin{bmatrix} F_{k,1} & \cdots & F_{k,P} \\ \vdots & & \vdots \\ F_{k+(K-1)M,1} & \cdots & F_{k+(K-1)M,P} \end{bmatrix}, \qquad (2.73)$$

where

$$\mathbf{F}_p(k) = \begin{bmatrix} F_{k,p} \\ F_{k+M,p} \\ \vdots \\ F_{k+(K-1)M,p} \end{bmatrix}. \qquad (2.74)$$

The optimization of the FDE coefficients under the MMSE criterion leads to better performances then the ZF criterion for the reasons that were already mentioned. Therefore, we will consider only a linear MMSE-FDE which can be regarded as a linear MUD under the MMSE criterion since the receiver need to detect all users.

To compute the FDE coefficients matrix, let us start by including (2.68) into (2.72), i.e.,

$$\begin{aligned} \tilde{\mathbf{A}}(k) &= \mathbf{F}^T(k)\mathbf{H}^T(k)\mathbf{A}(k) + \mathbf{F}^T(k)\mathbf{N}(k) \\ &= \underbrace{\mathbf{\Gamma}(k)\mathbf{A}(k)}_{\text{Useful signal}} + \underbrace{\left(\mathbf{F}^T(k)\mathbf{H}^T(k) - \mathbf{\Gamma}(k)\right)\mathbf{A}(k)}_{\text{MAI}} \\ &\quad + \underbrace{\mathbf{F}^T(k)\mathbf{N}(k)}_{\text{Channel noise}} \end{aligned} \qquad (2.75)$$

where

$$\mathbf{\Gamma}(k) = [\mathbf{\Gamma}_1(k) \ldots \mathbf{\Gamma}_P(k)] = \text{diag}(\gamma_1, \ldots, \gamma_P) \qquad (2.76)$$

with

$$\gamma_p = \frac{1}{M}\mathbf{F}_p^T(k)\mathbf{H}_p^T(k) = \frac{1}{M}\sum_{k\in\Psi_m}H_{k,p}F_{k,p}. \tag{2.77}$$

The coefficient $\gamma_p$ can be regarded as the average "overall" channel frequency response, for the $p$th user, after the filter $\mathbf{F}_p(k)$. By observing (2.75), we can identify three different terms: the first is the "useful" signal component, the second is the MAI and the last component is concerned with the channel noise. Note that, if there was no MAI at the output of the FDE, the "overall" channel frequency response $\mathbf{F}^T(k)\mathbf{H}^T(k)$ would be diagonal. Therefore, the MAI component in the frequency-domain is associated to the non-zero elements outside the diagonal of $\mathbf{F}^T(k)\mathbf{H}^T(k)$. The equalized samples vector $\tilde{\mathbf{A}}(k)$ can then be written as

$$\tilde{\mathbf{A}}(k) = \mathbf{\Gamma}(k)\mathbf{A}(k) + \mathbf{\Theta}(k) \tag{2.78}$$

where $\mathbf{\Gamma}(k) = \mathbf{I}_P$ and

$$\mathbf{\Theta}(k) = \tilde{\mathbf{A}}(k) - \mathbf{A}(k) = \begin{bmatrix} \Theta_{m,1} \\ \vdots \\ \Theta_{m,P} \end{bmatrix} \tag{2.79}$$

denotes the length-$P$ column vector with the "total" error in the $m$th data symbol of each user, that includes both the channel noise and MAI.

The FDE coefficients matrix $\mathbf{F}(k)$ are chosen so as to maximize the SINR for all users. For the $p$th user the SINR is defined as

$$\text{SINR}_p = \frac{E\left[|A_{m,p}|^2\right]}{E\left[|\Theta_{m,p}|^2\right]}. \tag{2.80}$$

Therefore, the maximization of $\{\text{SINR}_p, p = 1, \ldots, P\}$ is equivalent to the minimization of

$$E\left[|\mathbf{\Theta}(k)|^2\right] = E\left[\left|(\mathbf{F}^T(k)\mathbf{H}^T(k) - \mathbf{I}_P)\mathbf{A}(k)\right|^2\right] + E\left[\left|\mathbf{F}^T(k)\mathbf{N}(k)\right|^2\right] \tag{2.81}$$

conditioned to $\mathbf{\Gamma}(k) = \mathbf{I}_P$.

This minimization can be performed by employing the Lagrange multipliers' method. For this purpose, we can define the matrix of Lagrange functions

$$\mathbf{J} = E\left[|\mathbf{\Theta}(k)|^2\right] + (\mathbf{\Gamma}(k) - \mathbf{I}_P)\mathbf{\Lambda}, \tag{2.82}$$

where

$$\mathbf{\Lambda} = [\mathbf{\Lambda}_1 \dots \mathbf{\Lambda}_P] = \mathrm{diag}(\lambda_1, \lambda_2, \dots, \lambda_P) \qquad (2.83)$$

is the $P \times P$ diagonal matrix with the Lagrange multipliers, and assume that the optimization is carried out under $\mathbf{\Gamma}(k) = \mathbf{I}_P$. The optimum FDE coefficients matrix is obtained by solving the following set of equations:

- $\nabla_{\mathbf{F}(k)} J = 0 \Leftrightarrow \mathbf{H}^H(k)\mathbf{R}_A \mathbf{H}(k)\mathbf{F}(k) - \mathbf{H}^H(k)\mathbf{R}_A + \mathbf{R}_N \mathbf{F}(k)$

$$+ \frac{1}{M}\mathbf{H}^H(k)\mathbf{\Lambda} = 0$$

$$\Leftrightarrow \mathbf{H}^H(k)\mathbf{H}(k)\mathbf{F}(k) - \mathbf{H}^H(k) + \beta \mathbf{F}(k)$$

$$+ \frac{1}{2M\sigma_A^2}\mathbf{H}^H(k)\mathbf{\Lambda} = 0, \qquad (2.84)$$

with

$$\mathbf{R}_A = E\left[\mathbf{A}^*(k)\mathbf{A}^T(k)\right] = 2\sigma_A^2 \mathbf{I}_P, \qquad (2.85)$$

$$\mathbf{R}_N = E\left[\mathbf{N}^*(k)\mathbf{N}^T(k)\right] = 2\sigma_N^2 \mathbf{I}_K \qquad (2.86)$$

and

$$\beta = \frac{E\left[|N_k|^2\right]}{E\left[|A_{m,p}|^2\right]} = \frac{\sigma_N^2}{\sigma_A^2}, \qquad (2.87)$$

where $\sigma_N^2$ and $\sigma_A^2$ denote the variance of the noise term and the variance of the data symbols, respectively;

- $\nabla_{\mathbf{\Lambda}} J = 0 \Leftrightarrow \mathbf{\Gamma}(k) = \mathbf{I}_P.$ \hfill (2.88)

As expected, (2.88) is the condition under which the optimization is carried out.

Rewriting (2.84), the optimum FDE coefficients matrix is given by

$$\mathbf{F}(k) = \left[\mathbf{H}^H(k)\mathbf{H}(k) + \beta \mathbf{I}_K\right]^{-1}\mathbf{H}^H(k)\mathbf{Q}, \qquad (2.89)$$

where the normalization matrix

$$\mathbf{Q} = \mathrm{diag}(Q_1, \dots, Q_P) = \mathbf{I}_P - \frac{1}{2M\sigma_A^2}\mathbf{\Lambda} \qquad (2.90)$$

ensures that $\Gamma(k) = \mathbf{I}_P$.

Figure 2.25 shows an example of the average BER performance for an uplink transmission of an MC-CDMA system with a linear MUD receiver under the MMSE criterion. Each user transmits independent data blocks with length $N = KM = 256$ subcarriers and a spreading factor $K = 16$ (corresponding to $M = 16$ data symbols per block), plus an appropriate CP, over the same strong frequency-selective channel considered previously. The number of users is $P = 1, 4, 8$ and $16$ whose signals have the same average power (i.e., a perfect "average power control"). MFB performance and SU performance, defined as

$$P_{b,MFB} = E\left[ Q\left( \sqrt{\frac{2E_b}{N_0} \frac{1}{N} \sum_{k=0}^{N-1} \sum_{l=1}^{L} \left| H_{k,p}^{(l)} \right|^2} \right) \right], \tag{2.91}$$

and

$$P_{b,SU} = \frac{1}{M} \sum_{m=0}^{M-1} E\left[ Q\left( \sqrt{\frac{2E_b}{N_0} \frac{1}{K} \sum_{k \in \Psi_m} \sum_{l=1}^{L} \left| H_{k,p}^{(l)} \right|^2} \right) \right] \tag{2.92}$$

respectively, (again with the expectation taken over the set of channel realizations and $E[|H_{k,p}^{(l)}|^2] = 1$, for any $k$) are also included. Clearly, $P_{b,MFB} = P_{b,SU}$ when $K = N$; for $K < N$, $P_{b,SU}$ is typically worse than $P_{b,MFB}$. From Fig. 2.25 it is clear that the performance degradation increases as the number of users increases due to severe MAI levels resulting from the loss of orthogonality between the different spreading codes.

Another example is shown in Fig. 2.26 where different power are assigned to different users. In this case, there is a total of $P = K = 8$ users (i.e., a fully loaded system), half of them belonging to $C_L$ class and the other half belonging to $C_H$ class. $C_H$ users have an average power 6 dB above the average power of $C_L$ users. As expected, the performances are very poor, especially for $C_L$ users which face stronger MAI levels.

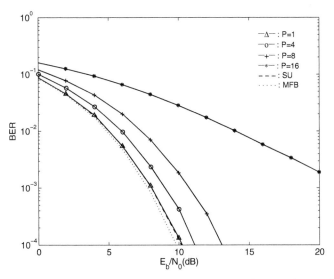

Figure 2.25  Average uncoded BER performance for a linear MUD under the MMSE criterion with $N = 256$ and $K = 16$ spreading codes.

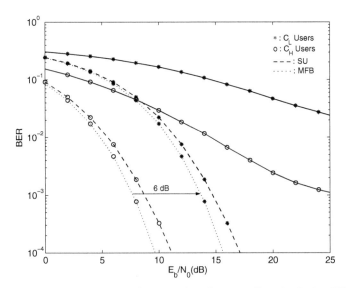

Figure 2.26  Average uncoded BER performance for a linear MUD under the MMSE criterion with $N = 256$ and a fully loaded scenario ($K = P = 8$), with four $C_L$ users and four $C_H$ users as a function of the $E_b/N_0$ of $C_H$ users.

## 2.3 DS-CDMA Systems with Linear FDE

As with MC-CDMA, in DS-CDMA systems all users transmit continuously, regardless of the bit rates. Therefore, the peak power requirements for the amplifiers are significantly reduced. Moreover, since DS-CDMA schemes can be regarded as single-carrier modulations, the transmitted signal associated to each spreading code can have low envelope fluctuations contrarily to MC-CDMA schemes. For these reasons, DS-CDMA schemes are good candidates for broadband wireless systems, especially at the uplink. DS-CDMA schemes can be combined with CP-assisted block transmission techniques, allowing a frequency-domain receiver design that have relatively low complexity, even for severely time-dispersive channels.

### 2.3.1 Downlink Transmission

Let us consider the downlink transmission in a DS-CDMA system. The BS simultaneously transmits independent blocks of $M$ data symbols for $P$ users, as depicted in Fig. 2.27. It is assumed that all users have the same spreading factor and the same rate. The transmitted block of time-domain chips is $\{s_n; n = 0, 1, \ldots, N - 1\}$, once again with $N = KM$.[6] The overall "chip" symbols $s_n$ are given by

$$s_n = \sum_{p=1}^{P} \xi_p s_{n,p},$$  (2.93)

where $\xi_p$ is the weighting coefficient for power control proposes, and

$$s_{n,p} = c_{n,p} a_{\lfloor n/K \rfloor, p}$$  (2.94)

is the $n$th chip for the $p$th user. The block $\{a_{m,p}; m = 0, 1, \ldots, M - 1\}$ denotes the time-domain data symbols associated to the $p$th user and $\{c_{n,p}; n = 0, 1, \ldots, N-1\}$ denotes the corresponding spreading sequence. It is assumed that the spreading sequences are periodic, with period $K$ (i.e., $c_{n+K,p} = c_{n,p}$, with $|c_{n,p}| = 1$) and results from the product of an $N$-length Walsh–Hadamard sequence with a pseudo-random QPSK sequence, common to all users of the BS.

The frequency-domain block $\{S_{k,p}; k = 0, 1, \ldots, N - 1\}$ is the DFT of the chip block associated to the $p$th user, $\{s_{n,p}; n = 0, 1, \ldots, N - 1\}$. Having

---

[6] Note that for MC-CDMA $N$ is the number of subcarriers, while for DS-CDMA $N$ is the number of chips per block.

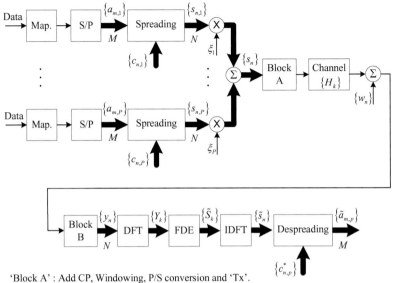

'Block A' : Add CP, Windowing, P/S conversion and 'Tx'.
'Block B' : 'Rx', S/P conversion, CP removal.

Figure 2.27  DS-CDMA downlink transmission model.

in mind (2.94), it can easily be shown that

$$S_{k,p} = A'_{k,p} C'_{k,p} \qquad (2.95)$$

where $\{A'_{k,p}; k = 0, 1, \ldots, N-1\}$ is the DFT of $\{a'_{n,p}; n = 0, 1, \ldots, N-1\}$, with

$$a'_{n,p} = \begin{cases} a_{m,p}, & n = mK \\ 0, & \text{otherwise} \end{cases} \qquad (2.96)$$

and $\{C'_{k,p}; k = 0, 1, \ldots, N-1\}$ is the DFT of $\{c'_{n,p}; n = 0, 1, \ldots, N-1\}$, with

$$c'_{n,p} = \begin{cases} c_{n,p}, & 0 \leq n < K-1 \\ 0, & \text{otherwise.} \end{cases} \qquad (2.97)$$

Clearly,

$$A'_{k,p} = \frac{1}{K} A_{k \bmod M, p}, \qquad (2.98)$$

$(k = 0, 1, \ldots, N-1)$ with $\{A_{k,p}; k = 0, 1, \ldots, M-1\}$ the DFT of $\{a_{m,p}; m = 0, 1, \ldots, M-1\}$. This means that, apart a constant, the block $\{A'_{k,p}; k = 0, 1, \ldots, N-1\}$ is the size-$N$ periodic extension of the DFT of the data block $\{A_{k,p}; k = 0, 1, \ldots, M-1\}$ associated to the $p$th user. This

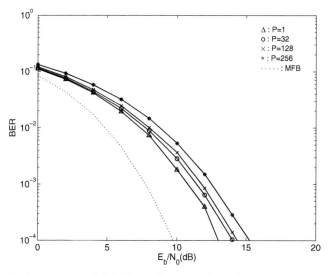

Figure 2.28 Average uncoded BER performance for a DS-CDMA transmission with an MMSE-FDE for $N = K = 256$ and $P = 1, 32, 128$ and $256$ users.

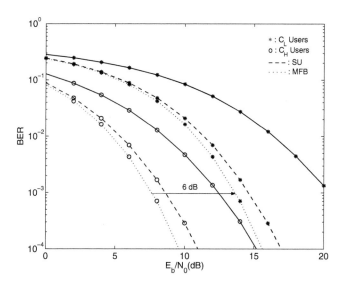

Figure 2.29 Average uncoded BER performance for a DS-CDMA transmission with $N = 256$ and a fully loaded scenario ($K = P = 8$) with four $C_L$ users and four $C_H$ users as a function of the $E_b/N_0$ of $C_H$ users.

multiplicity in the $A'_{k,p}$ is related to the spectral correlations that are inherent to the cyclostationary nature of the transmitted signals [45]. This means that there is a $K$-order implicit diversity effect in the received frequency-domain samples $\{Y_k; k = 0, 1, \ldots, N - 1\}$, as with MC-CDMA. Therefore, the receiver can be an FDE optimized under the MMSE criterion, followed by a suitable despreading procedure, leading to

$$\tilde{a}_{m,p} = \sum_{n'=mK}^{mK+K-1} \tilde{s}_{n'} c^*_{n',p}, \qquad (2.99)$$

where $\{\tilde{s}_n; n = 0, 1, \ldots, N - 1\}$ is the IDFT of $\{\tilde{S}_k; k = 0, 1, \ldots, N - 1\}$, with $\tilde{S}_k$ given by (2.52).

Figure 2.28 shows some performance results for a DS-CDMA downlink transmission with an MMSE-FDE for $N = K = 256$ and $P = 1, 32, 128$ and 256 users with the same average power. Figure 2.29 shows the average BER performance for two classes of users with different average power, as in Fig. 2.23. Observing these figures, similar conclusions can be taken as the ones concerning the MMSE-FDE of Figs. 2.22 and 2.23 for the downlink transmission of the MC-CDMA scheme: since the orthogonality of the spreading codes is lost due to the channel strong frequency selectivity and cannot be fully recovered with a MMSE-FDE, the performances remain far from the MFB performance, especially for fully loaded scenarios and/or when different power are assigned to different users.

### 2.3.2 Uplink Transmission

As with the uplink transmission of MC-CDMA schemes, in the uplink of a DS-CDMA scheme the signals associated to different users face different propagation channels resulting in the loss of orthogonality of the spreading codes in severe time-dispersive channel, requiring some kind of MUD.

Let us start by characterizing the uplink transmission of a DS-CDMA system involving $P$ users, as depicted in Fig. 2.30. The size-$M$ data symbol block to be transmitted by the $p$th user is $\{a_{m,p}; m = 0, 1, \ldots, M - 1\}$, with $a_{m,p}$ selected from a given constellation. The corresponding block of chips to be transmitted is $\{s_{n,p}; n = 0, 1, \ldots, N - 1\}$ with $s_{n,p}$ given by (2.94), as illustrated in Fig. 2.31. At the BS, the received frequency-domain block is $\{Y_k; k = 0, 1, \ldots, N - 1\}$, with $Y_k$ given by (2.64) or in matrix form by (2.68).

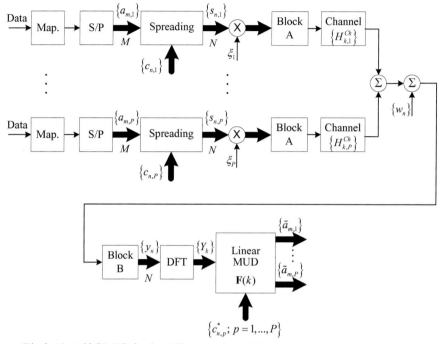

'Block A' : Add CP, Windowing, P/S conversion and 'Tx'.
'Block B' : 'Rx', S/P conversion, CP removal.

Figure 2.30 DS-CDMA uplink transmission model.

For a linear MUD under the MMSE criterion, the optimization of the FDE coefficients matrix $\mathbf{F}(k)$ can be carried out exactly as in the uplink of MC-CDMA scheme, described in Section 2.2.2, leading to the optimum solution, in the MMSE sense, given by (2.89). The detection of the data symbols $\{a_{m,p}; m = 0, 1, \ldots, M-1\}$ transmitted by each user can be estimated from the time-domain samples $\{\tilde{s}_n; n = 0, 1, \ldots, N-1\}$ corresponding to the IDFT of the equalized samples $\{\tilde{S}_k; k = 0, 1, \ldots, N-1\}$ using (2.99).

By observing Figs. 2.32 and 2.33, which show the average BER performance for the uplink of a DS-CDMA transmission with a linear MUD in similar conditions as the ones considered previously, we can again conclude that when the number of users increase and/or when different powers are assigned to different users the levels of interference between users lead to poor performance, still several dB far from the MFB performance.

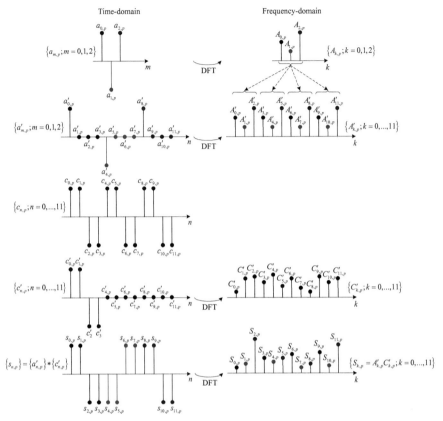

Figure 2.31 Illustration of the time-domain spreading and the multiplicity in $A'_{k,p}$ for $M = 3$ symbols and a spreading factor of $K = 4$.

## 2.4 Final Remarks

From the analysis presented in the last two sections, concerning both the downlink and uplink transmissions of MC-CDMA and DS-CDMA systems, it was clear that the use of linear frequency-domain receiver implementations are especially interesting for high data rate CP-assisted block transmissions over severely time-dispersive channels, since they allow quite efficient FFT-based implementations, with much lower complexity then the optimum receivers. However, when an MMSE-FDE is used instead of a ZF-FDE to avoid noise enhancement effects, severe residual interference levels can be expected. In fact, our performance results show that, for high system load and/or when different users have different assigned power, the high levels

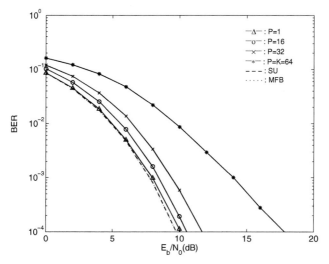

Figure 2.32 Average uncoded BER performance for a linear MUD with an MMSE-FDE with $N = 256$, $K = 64$ and $P = 1$, 16, 32 and 64 users.

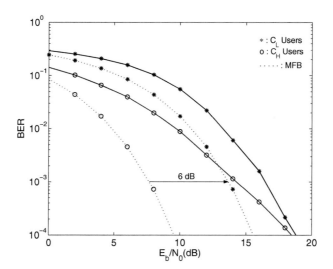

Figure 2.33 Average uncoded BER performance for a linear MUD with an MMSE-FDE with $N = 256$ and a fully loaded scenario ($K = P = 8$), with four $C_L$ users and four $C_H$ users as a function of the $E_b/N_0$ of $C_H$ users.

of interference invariably lead to BER performance substantially far from the corresponding MFB and SU performance in both downlink and uplink transmissions of MC-CDMA and DS-CDMA schemes.

The main objective of this work consists of designing more sophisticated frequency-domain receiver structures able to overcome these problems and significantly improve the performance in CDMA systems, without compromising too much the signal processing requirements of the receivers. To accomplish this goal we present an unified approach for MC-CDMA and DS-CDMA receivers, that considers the use of nonlinear equalizers based on the IB-DFE concept previously introduced in this chapter, whose potential was already pointed out in the SC-FDE context. It will be shown that by combining interference cancelation with MUD techniques good performance/complexity tradeoffs can be achieved, even for strongly time-dispersive channels.

# 3

# Receivers for the Downlink Transmissions of DS-CDMA and MC-CDMA Systems

It was shown in Chapter 2 that CP-assisted block transmission techniques employing FDE are suitable for high data rate transmission over severely time-dispersive channels. However, when linear MMSE-FDE techniques are employed within CP-assisted CDMA systems, the performance for the downlink transmission remains far from the corresponding MFB and SU performance, namely, due to significant residual interference levels, especially for fully loaded systems and/or when different powers are assigned to different spreading codes. ZF-FDE receivers can avoid residual interference levels but have the noise enhancement problems already mentioned.

To avoid these problems, several techniques were proposed to enhance the performance of MC-CDMA and DS-CDMA systems. In [46], the use of soft information from the channel decoder output for interference cancelation is exploited in MC-CDMA systems. A similar approach making use of soft information from the channel decoding stage to suppress MUI interference is the turbo receiver proposed in [47] for DS-CDMA systems.

This chapter deals with the receiver design for the downlink transmission of CP-assisted DS-CDMA and MC-CDMA systems. The receivers are based on the IB-DFE concept and extended herein to CDMA systems with space diversity, as an alternative to conventional linear FDE techniques.

The publications resulting from the work presented in this chapter are [48–51].

The chapter is organized as follows: The IB-DFE receiver techniques are addressed in Section 3.1, where their parameters are derived. Section 3.2 introduces "soft decisions" and the turbo equalization concept showing how turbo equalizers can be defined based on IB-DFE receivers. Finally, a set of performance results is presented and discussed in Section 3.3.

55

## 3.1 IB-DFE with Hard Decisions

This section describes the IB-DFE receiver techniques for both DS-CDMA and MC-CDMA systems. A detailed analysis on the receivers parameters and its derivation is also carried out.

### 3.1.1 Receiver Structure

Figure 3.1 presents the IB-DFE receiver structure with an $L$-branch space diversity for DS-CDMA and MC-CDMA schemes. In both cases, for a given iteration $i$, the output samples are given by (2.44), which is repeated here for convenience, i.e.,

$$\tilde{S}_k^{(i)} = \sum_{l=1}^{L} F_k^{(l,i)} Y_k^{(l)} - B_k^{(i)} \hat{S}_k^{(i-1)} \tag{3.1}$$

where $\{F_k^{(l,i)}; k = 0, 1, \ldots, N - 1\}$ $(l = 1, \ldots, L)$ and $\{B_k^{(i)}; k = 0, 1, \ldots, N - 1\}$ denote the feedforward and the feedback equalizer coefficients for the $i$th iteration, respectively, optimized so as to maximize the "overall" SINR, as described in the following. The block $\{\hat{S}_k^{(i-1)}; k = 0, 1, \ldots, N - 1\}$ is an estimate of the transmitted block $\{S_k^{(i-1)}; k = 0, 1, \ldots, N-1\}$, obtained from the data estimates of the $(i - 1)$th iteration, $\{\hat{A}_{m,p}^{(i-1)}; m = 0, 1, \ldots, M - 1\}$ in the MC-CDMA case and $\{\hat{a}_{m,p}^{(i-1)}; m = 0, 1, \ldots, M - 1\}$ in the DS-CDMA case, as in (2.53)–(2.55) or (2.93)–(2.94), respectively.

The data estimates are the "hard decisions" associated to the despread samples $\{\tilde{A}_{m,p}^{(i-1)}; m = 0, 1, \ldots, M - 1\}$ in the MC-CDMA case and $\{\tilde{a}_{m,p}^{(i-1)}; m = 0, 1, \ldots, M - 1\}$ in the DS-CDMA case, given by (2.56) and (2.99), respectively.

### 3.1.2 Derivation of the Receiver Parameters

As with IB-DFE receivers for SC transmissions briefly presented in Chapter 2, if there were no ISI at the output of the feedforward filter, the "overall" channel frequency response $\sum_{l=1}^{L} F_k^{(l,i)} H_k^{(l)}$ would be constant. Therefore, the ISI component in the frequency-domain is associated to the difference between the average channel frequency response after the feedforward filter, defined as

$$\gamma^{(i)} = \frac{1}{N} \sum_{k=0}^{N-1} \sum_{l=1}^{L} F_k^{(l,i)} H_k^{(l)}, \tag{3.2}$$

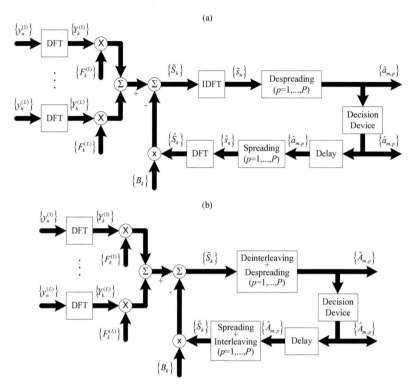

Figure 3.1 IB-DFE receiver structure with an $L$-branch space diversity for (a) DS-CDMA and (b) MC-CDMA.

and its actual value $\sum_{l=1}^{L} F_k^{(l,i)} H_k^{(l)}$. If we have reliable estimates of the transmitted block, the feedback filter can then be used to remove this residual ISI.

Therefore, the equalized frequency-domain samples associated to each iteration, $\tilde{S}_k^{(i)}$, can be written as

$$\tilde{S}_k^{(i)} = \gamma^{(i)} S_k + \Theta_k^{(i)}, \tag{3.3}$$

where $\Theta_k^{(i)} = \tilde{S}_k^{(i)} - \gamma^{(i)} S_k$ denotes an overall error that includes both the channel noise and the residual ISI. In the same way, the corresponding time-domain samples can be written as

$$\tilde{s}_n^{(i)} = \gamma^{(i)} s_n + \theta_n^{(i)} \tag{3.4}$$

where the block $\{\theta_n^{(i)}; n = 0, 1, \ldots, N-1\}$ is the IDFT of the block $\{\Theta_k^{(i)}; k = 0, 1, \ldots, N-1\}$.

The feedforward and feedback IB-DFE coefficients, $\{F_k^{(l,i)}; k = 0, 1, \ldots,$ $N - 1\}$ $(l = 1, \ldots, L)$ and $\{B_k^{(i)}; k = 0, 1, \ldots, N - 1\}$, respectively, are chosen so as to maximize the SINR, defined as

$$\text{SINR}^{(i)} = \frac{\left|\gamma^{(i)}\right|^2 E\left[|S_k|^2\right]}{E\left[\left|\Theta_k^{(i)}\right|^2\right]}. \tag{3.5}$$

The frequency-domain estimates, $\hat{S}_k^{(i)}$, can be written as [34]

$$\hat{S}_k^{(i)} = \rho^{(i)} S_k + \Delta_k^{(i)}, \tag{3.6}$$

where the correlation coefficient $\rho^{(i)}$, given by

$$\rho^{(i)} = \frac{E\left[\hat{s}_n^{(i)} s_n^*\right]}{E\left[|s_n|^2\right]} = \frac{E\left[\hat{S}_k^{(i)} S_k^*\right]}{E\left[|S_k|^2\right]}, \tag{3.7}$$

is a measure of the reliability of the decisions used in the feedback loop and $\Delta_k^{(i)}$ denotes a zero-mean error term. Since it is assumed that $E[\Delta_k^{(i)} S_{k'}^{(i)*}] \approx 0$ for $k' \neq k$,

$$E\left[\left|\Delta_k^{(i)}\right|^2\right] \approx \left(1 - \left(\rho^{(i)}\right)^2\right) E\left[|S_k|^2\right]. \tag{3.8}$$

The coefficient $\rho^{(i-1)}$, which can be regarded as the blockwise reliability of the decisions used in the feedback loop (from the previous iteration), is crucial for the good performance of the proposed receivers and can be estimated from the samples $\{\tilde{a}_{m,p}; m = 0, 1, \ldots, M - 1\}$ in the DS-CDMA case or $\{\tilde{A}_{m,p}; m = 0, 1, \ldots, M - 1\}$ in the MC-CDMA case, as described in the next subsection.

By combining (2.34), (3.1) and (3.6), we obtain

$$\tilde{S}_k^{(i)} = \sum_{l=1}^{L} F_k^{(l,i)} \left(H_k^{(l)} S_k + N_k^{(l)}\right) - B_k^{(i)} \left(\rho^{(i-1)} S_k + \Delta_k^{(i-1)}\right)$$

$$= \underbrace{\gamma^{(i)} S_k}_{\text{Useful signal}} + \underbrace{\left(\sum_{l=1}^{L} F_k^{(l,i)} H_k^{(l)} - \gamma^{(i)} - \rho^{(i-1)} B_k^{(i)}\right) S_k}_{\text{Residual Interference (ISI+MUI)}}$$

$$\underbrace{- B_k^{(i)} \Delta_k^{(i-1)}}_{\substack{\text{``Noise'' due to} \\ \text{feedback errors}}} + \underbrace{\sum_{l=1}^{L} F_k^{(l,i)} N_k^{(l)}}_{\text{Channel noise}}. \tag{3.9}$$

This means that $\tilde{S}_k^{(i)}$ has four terms: a "signal" component, $\gamma^{(i)}S_k$, and three "noise" components. The first component in the last equality of (3.9) is the residual ISI and "inter-code" interference; the second component accounts for the errors in $\hat{s}_n^{(i-1)}$; and the final component is concerned to the channel noise.

The maximization of the SINR in (3.5) is equivalent to the minimization of

$$
E\left[\left|\varepsilon_k^{Eq(i)}\right|^2\right] = E\left[\left|\sum_{l=1}^{L} F_k^{(l,i)} H_k^{(l)} - \gamma^{(i)} - \rho^{(i-1)} B_k^{(i)}\right|^2\right] E\left[|S_k|^2\right]
$$

$$
+ E\left[\left|B_k^{(i)} \Delta_k^{(i-1)}\right|^2\right] + E\left[\left|\sum_{l=1}^{L} F_k^{(l,i)} N_k^{(l)}\right|^2\right]
$$

$$
= E\left[\left|\sum_{l=1}^{L} F_k^{(l,i)} H_k^{(l)} - \gamma^{(i)} - \rho^{(i-1)} B_k^{(i)}\right|^2\right] 2\sigma_S^2
$$

$$
+ E\left[\left|B_k^{(i)}\right|^2\right]\left(1 - \left(\rho^{(i-1)}\right)^2\right) 2\sigma_S^2
$$

$$
+ \sum_{l=1}^{L} E\left[\left|F_k^{(l,i)}\right|^2\right] 2\sigma_N^2, \tag{3.10}
$$

conditioned to a given $\gamma^{(i)}$, where $2\sigma_S^2 = E[|S_k|^2]$. For a normalized FDE, assuming that the optimization is carried out under $\gamma^{(i)} = 1$, the optimum receiver coefficients can be obtained by employing the Lagrange multipliers' method. For this purpose, we can define the function

$$
J = E\left[\left|\varepsilon_k^{Eq(i)}\right|^2\right] + \lambda^{(i)}\left(\frac{1}{N}\sum_{k'=0}^{N-1}\sum_{l=1}^{L} F_{k'}^{(l,i)} H_{k'}^{(l)} - 1\right), \tag{3.11}
$$

with $\lambda^{(i)}$ denoting the Lagrange multipliers. The optimum receiver coefficients are obtained by solving the following set of $L + 2$ equations:

$$
\bullet \quad \frac{\partial J}{\partial F_k^{(l,i)}} = 4\sigma_S^2 H_k^{(l)*}\left(\sum_{l'=1}^{L} F_k^{(l',i)} H_k^{(l')} - 1 - \rho^{(i-1)} B_k^{(i)} + \frac{\lambda^{(i)}}{2\sigma_S^2 N}\right)
$$

$$
+ 4\sigma_N^2 F_k^{(l,i)} = 0, \quad l = 1, 2, \ldots, L; \tag{3.12}
$$

- $$\frac{\partial J}{\partial B_k^{(i)}} = -4\sigma_S^2 \rho^{(i-1)} \left( \sum_{l'=1}^{L} F_k^{(l',i)} H_k^{(l')} - 1 - \rho^{(i-1)} B_k^{(i)} \right)$$
$$+ 4\sigma_S^2 \left( 1 - \left( \rho^{(i-1)} \right)^2 \right) B_k^{(i)} = 0; \qquad (3.13)$$

- $$\frac{\partial J}{\partial \lambda^{(i)}} = \frac{1}{N} \sum_{k=0}^{N-1} \sum_{l=1}^{L} F_k^{(l,i)} H_k^{(l)} - 1 = 0. \qquad (3.14)$$

As expected, (3.14) is equivalent to $\gamma^{(i)} = 1$. Equations (3.12) and (3.13) can be rewritten in the form

$$H_k^{(l)*} \left( \sum_{l'=1}^{L} F_k^{(l',i)} H_k^{(l')} - 1 - \rho^{(i-1)} B_k^{(i)} + \frac{\lambda^{(i)}}{2\sigma_S^2 N} \right) + \beta F_k^{(l',i)} = 0, \quad (3.15)$$

$l = 1, 2, \ldots, L$, and

$$\rho^{(i-1)} \left( \sum_{l'=1}^{L} F_k^{(l',i)} H_k^{(l')} - 1 - \rho^{(i-1)} B_k^{(i)} \right) = \left( 1 - \left( \rho^{(i-1)} \right)^2 \right) B_k^{(i)}, \quad (3.16)$$

where

$$\beta = \frac{E\left[ |N_k|^2 \right]}{E\left[ |S_k|^2 \right]} = \frac{\sigma_N^2}{\sigma_S^2}. \qquad (3.17)$$

From (3.16), the optimum values of $B_k^{(i)}$ are

$$B_k^{(i)} = \rho^{(i-1)} \left( \sum_{l'=1}^{L} F_k^{(l',i)} H_k^{(l')} - 1 \right). \qquad (3.18)$$

By replacing (3.18) in (3.15), we get the set of $L$ equations

$$\left( 1 - \left( \rho^{(i-1)} \right)^2 \right) H_k^{(l)*} \sum_{l'=1}^{L} F_k^{(l',i)} H_k^{(l')} + \beta F_k^{(l,i)}$$

$$= \left( 1 - \left( \rho^{(i-1)} \right)^2 - \frac{\lambda^{(i)}}{2\sigma_S^2 N} \right) H_k^{(l)*}, \qquad (3.19)$$

$l = 1, 2, \ldots, L.$

It can be easily verified by substitution that the solutions of (3.19) are

$$F_k^{(l,i)} = \frac{\mathcal{K}_F^{(i)} H_k^{(l)*}}{\beta + \left(1 - \left(\rho^{(i-1)}\right)^2\right) \sum_{l'=1}^{L} \left|H_k^{(l')}\right|^2}, \tag{3.20}$$

$l = 1, 2, \ldots, L$, where the normalization constant

$$\mathcal{K}_F^{(i)} = 1 - \left(\rho^{(i-1)}\right)^2 - \frac{\lambda^{(i)}}{2\sigma_S^2 N} \tag{3.21}$$

ensures that $\gamma^{(i)} = 1$. These feedforward coefficients can be used in (3.18) for obtaining the feedback coefficients $B_k^{(i)}$.

Clearly, for the first iteration ($i = 1$), no information exists about $\{S_k; k = 0, 1, \ldots, N - 1\}$ and the correlation coefficient in (3.20) is zero. This means that $B_k^{(1)} = 0$ and

$$F_k^{(l,1)} = \frac{\mathcal{K}_F^{(1)} H_k^{(l)*}}{\beta + \sum_{l'=1}^{L} \left|H_k^{(l')}\right|^2}, \tag{3.22}$$

$l = 1, 2, \ldots, L$, corresponding to the optimum frequency-domain equalizer coefficients under the MMSE criterion [22, 26]. After the first iteration, and if the residual BER is not too high (at least for the spreading codes with higher transmit power), we can use the feedback coefficients to eliminate a significant part of the residual interference. When $\rho \approx 1$ (after several iterations and/or for moderate-to-high SNRs), we have an almost full cancelation of the "inter-code" interference through these coefficients, while the feedforward coefficients perform an approximate matched filtering. Figure 3.2 illustrates the convergence of the despreaded samples $\{\tilde{A}_{m,p}; m = 0, 1, \ldots, M - 1\}$ with four iterations.

It should be pointed out that, when $L = 1$ (no diversity) the IB-DFE parameters derived above become identical to those given in [10]. It should also be noted that the feedforward coefficients can take the form

$$F_k^{(l,i)} = H_k^{(l)*} V_k^{(i)}, \tag{3.23}$$

$l = 1, 2, \ldots, L$, with

$$V_k^{(i)} = \frac{\mathcal{K}_F^{(i)}}{\beta + \left(1 - \left(\rho^{(i-1)}\right)^2\right) \sum_{l'=1}^{L} \left|H_k^{(l')}\right|^2}. \tag{3.24}$$

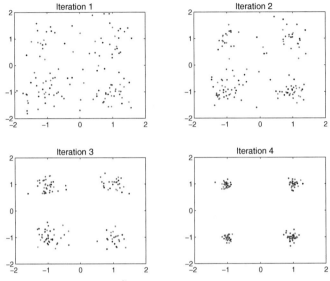

Figure 3.2 $\tilde{A}_{m,p}$ for iterations 1, 2, 3 and 4.

This means that the bank of feedforward filters can be replaced by a bank of matched filters which implement an ideal MRC, followed by a single feedforward filter characterized by the set of coefficients $\{V_k^{(i)}; k = 0, 1, \ldots, N-1\}$, as depicted in Fig. 3.3. For $\rho \approx 1$ this corresponds to a MRC receiver since the feedforward coefficients eliminate almost all ISI.

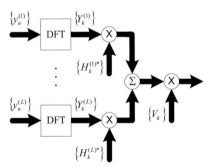

Figure 3.3 Equivalent FDE receiver structure with $L$-branch space diversity.

### 3.1.3 Calculation of $\rho_p$

In this subsection it is shown how we can obtain an estimate of the correlation coefficient. Assuming uncorrelated data blocks, it can be easily shown that

$$\rho^{(i-1)} = \sum_{p=1}^{P} \xi_p^2 \rho_p^{(i-1)}, \tag{3.25}$$

with

$$\rho_p^{(i-1)} = \frac{E\left[\hat{A}_{m,p} A_{m,p}^*\right]}{E\left[|A_{m,p}|^2\right]} = \frac{E\left[\hat{a}_{m,p} a_{m,p}^*\right]}{E\left[|a_{m,p}|^2\right]}, \tag{3.26}$$

denoting the correlation coefficient associated to the $p$th user. For a DS-CDMA scheme $\rho_p^{(i)}$ can be obtained as follows (a similar approach could be employed for MC-CDMA schemes).

Let us assume that the transmitted symbols $\{a_{m,p}; m = 0, 1, \ldots, M - 1\}$ belong to a QPSK constellation (the generalization to other constellations is straightforward). In this case,

$$a_{m,p} = a_{m,p}^I + j a_{m,p}^Q = \pm d \pm jd \tag{3.27}$$

where

$$a_{m,p}^I = \text{Re}\{a_{m,p}\} \tag{3.28}$$

and

$$a_{m,p}^Q = \text{Im}\{a_{m,p}\} \tag{3.29}$$

are the "in-phase" and "quadrature" components of $a_{m,p}$, respectively, and $d = D/2$, with $D$ corresponding to the minimum Euclidean distance (for the sake of simplicity, in the following we will ignore the dependency with the iteration number $i$). In this case,

$$E\left[|a_{m,p}|^2\right] = \frac{D^2}{4}. \tag{3.30}$$

For an unbiased FDE ($\gamma = 1$), the time-domain samples at the output of the FDE are

$$\tilde{a}_{m,p} = \tilde{a}_{m,p}^I + j\tilde{a}_{m,p}^Q = a_{m,p} + \theta_{m,p}, \tag{3.31}$$

where

$$\tilde{a}_{m,p}^I = \text{Re}\{\tilde{a}_{m,p}\}, \tag{3.32}$$

$$\tilde{a}_{m,p}^Q = \text{Im}\{\tilde{a}_{m,p}\} \tag{3.33}$$

and $\theta_{m,p}$ is the "overall" error component. It is assumed that $\theta_{m,p}$ is approximately Gaussian-distributed,[1] with $E[\theta_{m,p}] = 0$. Moreover, the SNR for detection purposes is

$$\text{SINR}_p = \frac{E\left[|a_{m,p}|^2\right]}{E\left[|\theta_{m,p}|^2\right]} = \frac{P|\xi_p|^2}{\displaystyle\sum_{p'=1}^{P} |\xi_{p'}|^2}\text{SINR}, \tag{3.34}$$

with SINR given by (3.5), i.e., $\text{SINR}_p$ is higher for the users with higher assigned power.

The symbol estimates can be written as

$$\hat{a}_{m,p} = a_{m,p} + v^I_{m,p} + jv^Q_{m,p} \tag{3.35}$$

where the error coefficients $v^I_{m,p}$ (or $v^Q_{m,p}$) are zero if there is no error in $a^I_{m,p}$ (or $a^Q_{m,p}$) and $\pm D$ otherwise. This means that $v^I_{m,p}$ and $v^Q_{m,p}$ are random variables, both taking the values 0 and $\pm D$ with probabilities $1 - P_{e,p}$ and $P_{e,p}$, respectively. Therefore,

$$\rho_p = 1 - 2P_{e,p}, \tag{3.36}$$

where $P_{e,p}$ denotes the BER associated to the $p$th user, which can be approximated by

$$P_{e,p} \approx Q\left(\sqrt{\text{SINR}_p}\right), \tag{3.37}$$

for QPSK constellations.

## 3.2 IB-DFE with Soft Decisions

In this section it is showed how we can design efficient turbo FDE receivers based on the IB-DFE concept.

Equation (3.1) could be written as

$$\tilde{S}^{(i)}_k = \sum_{l=1}^{L} F^{(l,i)}_k Y^{(l)}_k - B^{(i)}_k \overline{S}^{(i-1)}_k, \tag{3.38}$$

with

$$\overline{S}^{(i-1)}_k = \rho^{(i-1)} \hat{S}^{(i-1)}_k. \tag{3.39}$$

---

[1] This assumption is reasonable under severely time-dispersive channel conditions.

Since $\rho^{(i-1)}$ can be regarded as the blockwise reliability of the estimates $\hat{S}_k^{(i-1)}$, $\overline{S}_k^{(i-1)}$ is the overall block average of $S_k^{(i-1)}$ conditioned to the FDE output. To improve the performances, the "blockwise averages" could be replaced by "symbol averages", as described in the following.

Let us consider an MC-CDMA scheme (again, a similar approach could be employed for DS-CDMA schemes) and assume that the transmitted symbols are selected from a QPSK constellation under a Gray mapping rule (the generalization to other cases is straightforward), i.e.,

$$A_{m,p} = \pm 1 \pm j = A_{m,p}^I + j A_{m,p}^Q, \tag{3.40}$$

with

$$A_{m,p}^I = \mathrm{Re}\{A_{m,p}\} \tag{3.41}$$

and

$$A_{m,p}^Q = \mathrm{Im}\{A_{m,p}\}, \tag{3.42}$$

$m = 0, 1, \ldots, M - 1$ and $p = 1, 2, \ldots, P$, (similar definitions can be made for $\tilde{A}_{m,p}$, $\hat{A}_{m,p}$ and $\overline{A}_{m,p}$).

The loglikelihood ratios (LLR) of the "in-phase bit" and the "quadrature bit", associated to $A_{m,p}^I$ and $A_{m,p}^Q$, respectively, are given by (see Appendix C)

$$L_{m,p}^I = \frac{2}{\sigma_p^2} \tilde{A}_{m,p}^I \tag{3.43}$$

and

$$L_{m,p}^Q = \frac{2}{\sigma_p^2} \tilde{A}_{m,p}^Q, \tag{3.44}$$

respectively, where

$$\sigma_p^2 = \frac{1}{2} E\left[\left|A_{m,p} - \tilde{A}_{m,p}\right|^2\right] \approx \frac{1}{2M} \sum_{m=0}^{M-1} \left|\hat{A}_{m,p} - \tilde{A}_{m,p}\right|^2 \tag{3.45}$$

denote the variance of the data symbols estimates associated to the $p$th spreading code.

Under a Gaussian assumption, the mean value of $A_{m,p}$ is

$$\overline{A}_{m,p} = \tanh\left(\frac{L_{m,p}^I}{2}\right) + j \tanh\left(\frac{L_{m,p}^Q}{2}\right)$$

$$= \rho_{m,p}^I \hat{A}_{m,p}^I + j \rho_{m,p}^Q \hat{A}_{m,p}^Q, \tag{3.46}$$

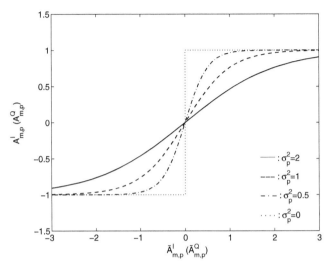

Figure 3.4  Soft decision function for different values of the data symbols' variances.

where the hard decisions $\hat{A}^I_{m,p} = \pm 1$ and $\hat{A}^Q_{m,p} = \pm 1$ are defined according to the signs of $L^I_{m,p}$ and $L^Q_{m,p}$, respectively, and $\rho^I_{m,p}$ and $\rho^Q_{m,p}$ can be regarded as the reliabilities associated to the "in-phase" and "quadrature" bits of the $m$th symbol of the $p$th spreading code, given by

$$\rho^I_{m,p} = \frac{E\left[A^I_{m,p}\hat{A}^I_{m,p}\right]}{E\left[\left|A^I_{m,p}\right|^2\right]} = \tanh\left(\frac{\left|L^I_{m,p}\right|}{2}\right) \tag{3.47}$$

and

$$\rho^Q_{m,p} = \frac{E\left[A^Q_{m,p}\hat{A}^Q_{m,p}\right]}{E\left[\left|A^Q_{m,p}\right|^2\right]} = \tanh\left(\frac{\left|L^Q_{m,p}\right|}{2}\right), \tag{3.48}$$

respectively (naturally, $0 \leq \rho^I_{m,p} \leq 1$ and $0 \leq \rho^Q_{m,p} \leq 1$). For the first iteration, $\rho^I_{m,p} = \rho^Q_{m,p} = 0$ and $\overline{A}_{m,p} = 0$; after some iterations and/or when the SNR is high, typically $\rho^I_{m,p} \approx 1$ and $\rho^Q_{m,p} \approx 1$, leading to $\overline{A}_{m,p} \approx \hat{A}_{m,p}$. Figure 3.4 shows the soft decision function (3.46) for different values of $\sigma^2_p$ and Fig. 3.5 compares the values of $\tilde{A}_{m,p}$, $\overline{A}_{m,p}$ and $\hat{A}_{m,p}$ at a given iteration.

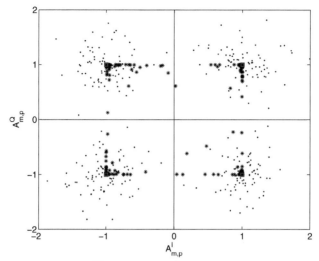

Figure 3.5 $\tilde{A}_{m,p}$ (•), $\overline{A}_{m,p}$ (∗) and $\hat{A}_{m,p}$ (□) at a given iteration.

The feedforward coefficients are still obtained from (3.20), with the blockwise reliability given by (3.25) and $\rho_p^{(i-1)}$ given by

$$\rho_p^{(i-1)} = \frac{1}{M} \sum_{m=0}^{M-1} \frac{E\left[\hat{A}_{m,p} A_{m,p}^*\right]}{E\left[\left|A_{m,p}\right|^2\right]} = \frac{1}{2M} \sum_{m=0}^{M-1} \left(\rho_{m,p}^{I^{(i-1)}} + \rho_{m,p}^{Q^{(i-1)}}\right). \quad (3.49)$$

The "overall chip averages" are then given by

$$\overline{S}_k = \sum_{p=1}^{P} \xi_p C_{k,p} \overline{A}_{\lfloor k/K \rfloor, p}. \quad (3.50)$$

Therefore, the receiver with "blockwise reliabilities", denoted in the following as IB-DFE with hard decisions, and the receiver with "symbol reliabilities", denoted in the following as IB-DFE with soft decisions, employ the same feedforward coefficients. However, in the first the feedback loop uses the "hard-decisions" on each data block, weighted by a common reliability factor, while in the second the reliability factor changes from symbol to symbol (in fact, the reliability factor is different in the real and imaginary component of each symbol).

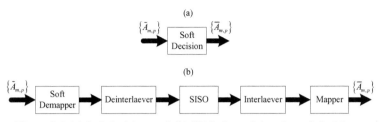

Figure 3.6 (a) Soft decisions and (b) SISO channel decoder soft decisions.

### 3.2.1 Turbo Equalizers

Most digital transmission schemes employ some sort of channel coding (as shown in Chapter 2, this can be particularly important for MC schemes such as OFDM) and the typical detection strategy is to perform separately the channel equalization and channel decoding procedures. However, it is known that high performance gains can be achieved if these procedures are jointly performed. An effective way of achieving this is by employing the so-called turbo equalization schemes where the equalization and decoding procedures are repeated, in an iterative way, with some soft information being passed between them [36]. Although initially proposed for time-domain receivers, turbo equalizers also allow frequency-domain implementations [52, 53].

It is also possible to define a turbo FDE receiver based on IB-DFE receiver that, as conventional turbo equalizers, employs the "soft decisions" from the channel decoder outputs (instead of the "soft decisions" from the FDE outputs) in the feedback loop. The receiver structure, is similar to the IB-DFE with soft decisions, but with a SISO channel decoder (Soft-In, Soft-Out) employed in the feedback loop, as in Fig. 3.6. The SISO block, that can be implemented as defined in [54], provides the LLRs of both the "information bits" and the "coded bits". The input of the SISO block are LLRs of the "coded bits" at the deinterleaver output. Once again, the feedforward coefficients are obtained from (3.20), with the blockwise reliability given by (3.49).

## 3.3 Performance Results

This section presents a set of performance results concerning the proposed IB-DFE receiver structures for the downlink transmission of DS-CDMA and MC-CDMA systems. It is assumed that each spreading code is assigned to a given user, with $N = 256$ subcarriers (similar results could be obtained for

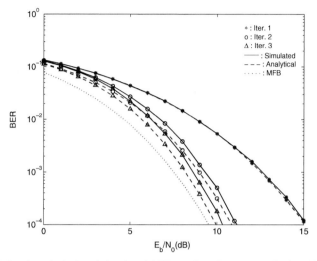

Figure 3.7 Semi-analytical and simulated BER results when $L = 1$, for iterations 1 to 3.

other values of $N$, provided that $N \gg 1$). An orthogonal spreading (with Walsh–Hadamard sequences) plus a pseudo-random scrambling (with the same chip rate) is considered for both DS-CDMA and MC-CDMA schemes. The strong frequency-selective channel described in Section 1.3 is considered and perfect synchronization and channel estimation are assumed. The number of users is $P = K$, i.e., we are assuming a fully loaded system. For the sake of comparisons, the MFB and SU performance given by (2.91) and (2.92), respectively, are also included.

Let us first assume that all users have the same power. In Fig. 3.7 we compare semi-analytical BER values, given by (3.37), with simulated ones for a DS-CDMA system with an IB-DFE with hard decisions, $N = K = 256$ and no space diversity ($L = 1$) (similar behaviors are observed for MC-CDMA systems). Clearly, the semi-analytical BER values are very close to the simulated ones for the first iteration; for the remaining iterations, the theoretical values are slightly optimistic (for $L > 1$ the semi-analytical BER values are even closer to the simulated ones). Figure 3.8 shows the evolution of the correlation factor $\rho$, together with the corresponding estimates (given by (3.36)), using an estimated BER obtained from the SNIR, as in (3.37). Clearly, the $\rho$ estimates are very close to the true $\rho$ values for the second iteration. When the number of iterations is increased, $\rho$ becomes slightly overestimated when the noise levels are high. For moderate-to-low noise levels, the $\rho$ estimates are still very accurate. The high accuracy of the $\rho$

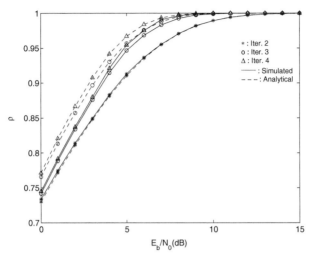

Figure 3.8  Evolution of $\rho$, when $L = 1$ (semi-analytical and simulated results).

estimates is a consequence of the approximated BER values given by (3.37) being close to the true ones (see also Fig. 3.7). We would like to point out that our receivers have good robustness relatively to errors in $\rho$, especially for $\hat{\rho} < \rho$, with the main problem being a slower convergence of the iterative process; for $\hat{\rho} > \rho$ there is slight performance degradation.

Figures 3.9 and 3.10 concern MC-CDMA and DS-CDMA systems, respectively, once again with $N = K = 256$ and $L = 1$ and 2 branch space diversity. Clearly, the iterative procedure allows a significant improvement relatively to the conventional linear FDE (first iteration). Moreover, the achievable performance is close to the MFB after three iterations. It is also observed that the performance is similar for MC-CDMA and DS-CDMA schemes. This is not surprising, since for $K = N$ all the available bandwidth is used to transmit each data symbols in both cases.

From Figs. 3.11 and 3.12, we can also conclude that the performance for MC-CDMA and DS-CDMA systems, respectively, is just slightly improved (a few tens of a dB, at most, in the MC-CDMA case) when an IB-DFE with soft decisions is used instead of the IB-DFE with hard decisions.

Let us consider now that $K = P = 256$ and two classes of users, $C_L$ and $C_H$, with $K/2 = 128$ users in each class, where the average power of $C_L$ users is 6 dB below the average power of $C_H$ users. Clearly, $C_L$ users face strong interference levels. Figures 3.13 and 3.14 present the BER performance for a MC-CDMA and DS-CDMA systems, respectively, expressed as

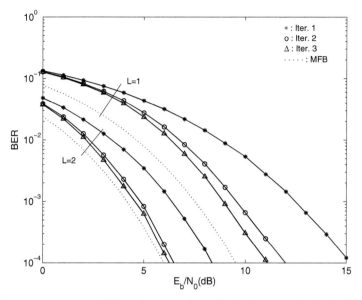

Figure 3.9 Average uncoded BER performance for MC-CDMA with $K = 256$ ($M = 1$) and $P = 256$ users, with the same assigned power.

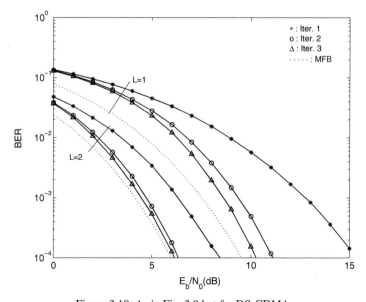

Figure 3.10 As in Fig. 3.9 but for DS-CDMA.

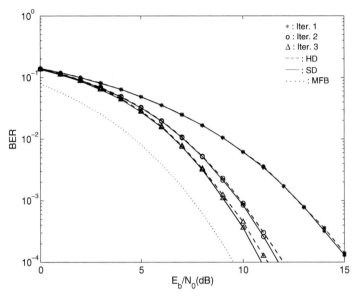

Figure 3.11 Average uncoded BER performance for MC-CDMA with $N = K = P = 256$ when hard decisions (HD) and soft decision (SD) are employed in the feedback loop.

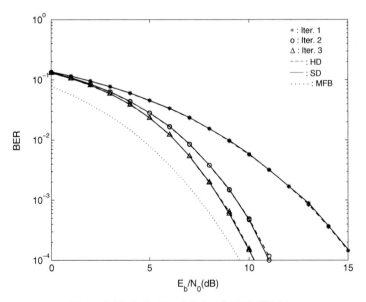

Figure 3.12  As in Fig. 3.11 but for DS-CDMA.

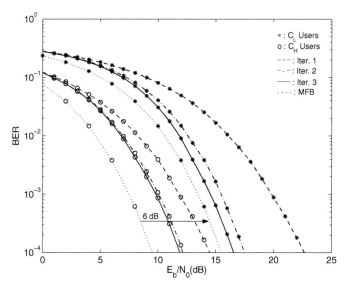

Figure 3.13 Average uncoded BER performance for MC-CDMA with $K/2 = 128 \, C_L$ users and $K/2 = 128 \, C_H$ users.

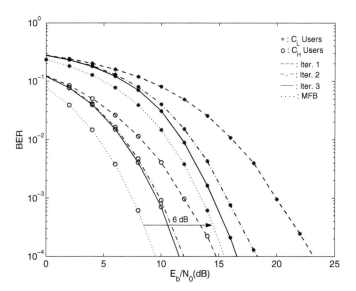

Figure 3.14 As in Fig. 3.13 but for DS-CDMA.

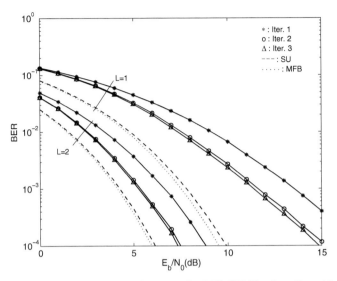

Figure 3.15 Average uncoded BER performance for MC-CDMA when $K = 16$ ($M = 16$) and $P = 16$ users, with the same assigned power.

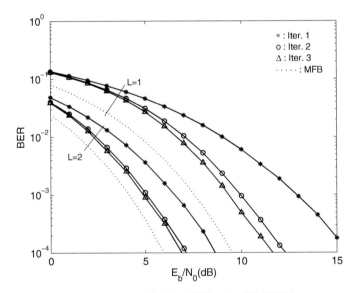

Figure 3.16 As in Fig. 3.15 but for DS-CDMA.

a function of the $E_b/N_0$ of $C_H$ users (6 dB below the $E_b/N_0$ of $C_L$ users). Once again, the iterative receiver allows significant performance improvements. From these figures, it is clear that performance gains associated to the iterative procedure are higher for $C_L$ users and the corresponding BERs are closer to the MFB than for $C_H$ users (the performance of $C_H$ users are still a few dB from the MFB after three iterations). This is explained as follows: the BER is much lower for $C_H$ users, allowing an almost perfect interference cancelation of their effects on $C_L$ users; the higher BERs for the $C_L$ users preclude an appropriate interference cancelation when $C_H$ users are detected.

It should also be noted that, for $K < N$, the performance of MC-CDMA schemes is worse, since just a fraction $1/M$ of the frequencies is used for the transmission of a given data symbol. This is not the case of DS-CDMA, where all frequencies can be used for transmitting each data symbol, regardless of the spreading factor. For instance, Figs. 3.15 and 3.16 concern the case where $K = 16$ (i.e., $M = 16$), with the same power is assigned to all spreading codes and $P = 16$ users (i.e., a fully loaded system) for MC-CDMA and DS-CDMA systems, respectively. From Fig. 3.15 we can observe that, although the iterative procedure allows gains of about 2 dB, the achievable performance is similar with two or three iteration, and still far from the MFB and the SU performance (the SU performance is slightly worse that the MFB when $K < N$). This is not true for the DS-CDMA scheme (see Fig. 3.16). However, it should be noted that this does not mean necessarily an weakness of the MC-CDMA schemes with small spreading factors. The comparison between DS-CDMA and MC-CDMA schemes should take into account other aspects, such as the envelope fluctuations of the transmitted signals and the impact of the channel coding (one might expect larger coding gains for MC-CDMA schemes, especially when a small $K$ is combined with interblock interleaving).

Let us consider now the impact of the channel coding by using a turbo FDE based on IB-DFE receivers. Again, the rate-1/2 64-state convolutional code described in Section 1.3 is used. An appropriate interblock interleaving of the coded bits is assumed before the mapping procedure (see Fig. 3.6). Figures 3.17 and 3.18 show the average coded BER performance for iterations 1 to 3, again for $L = 1$ or 2, for MC-CDMA and DS-CDMA schemes, respectively. As expected, the channel coding leads to significant performance improvements relatively to the uncoded case, especially for MC-CDMA scheme which performance approaches the MFB, confirming that channel coding (combined with appropriate interleaving) can compensate the worse uncoded performance.

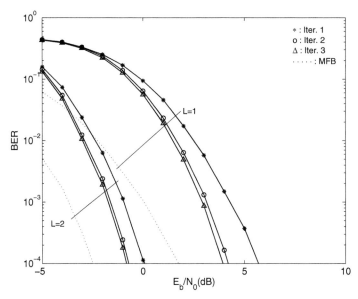

Figure 3.17  Average coded BER performance for MC-CDMA when $N = 256$ and $P = K = 16$ users, with the same assigned power.

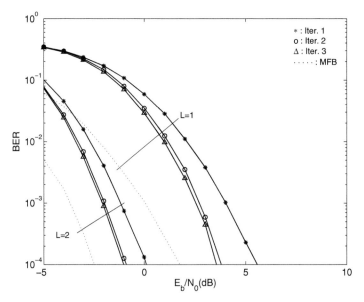

Figure 3.18  As in Fig. 3.17 but for DS-CDMA.

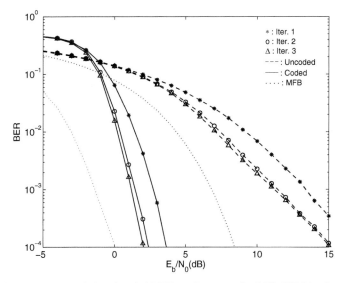

Figure 3.19 Average uncoded and coded BER performance for MC-CDMA when $N = 256$ and $P = K = 16$ users for a channel with uncorrelated Rayleigh fading on the different frequencies.

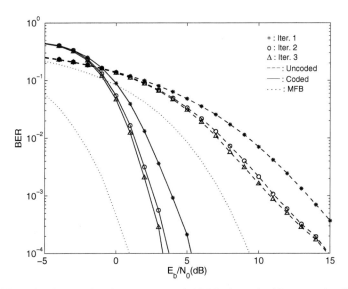

Figure 3.20 As in Fig. 3.19 but for an exponential PDP channel with uncorrelated Rayleigh fading on the different paths.

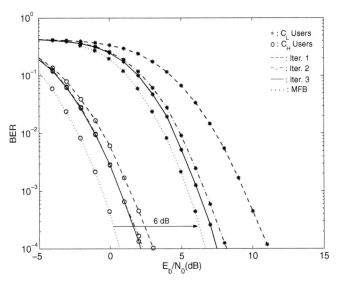

Figure 3.21  Average coded BER performance for DS-CDMA with $K/2 = 128$ $C_L$ users and $K/2 = 128$ $C_H$ users (average power of $C_L$ users 6 dB below the average power of $C_H$ users).

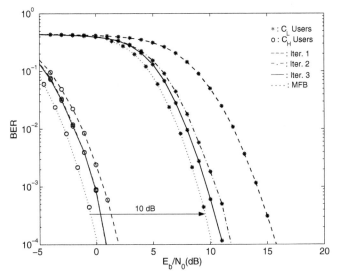

Figure 3.22  As in Fig. 3.21 but with the average power of $C_L$ users 10 dB below the average power of $C_H$ users.

Similar conclusions can be derived for other severely time-dispersive channels. For instance, Fig. 3.19 and 3.20 show the uncoded and coded BER performance for the MC-CDMA system (similar results are observed for the DS-CDMA system) for the same scenario of Figs. 3.15 and 3.17 (with $L = 1$) but with two different severely time-dispersive channels. Figure 3.19 concerns a channel with uncorrelated Rayleigh fading on the different frequencies and Fig. 3.20 concerns an exponential PDP channel with uncorrelated Rayleigh fading on the different paths (see Fig. A.2).

Figures 3.21 and 3.22 show the coded BER performance for DS-CDMA scheme for the scenario where different users have different assigned powers, when $C_L$ users have average assigned power 6 dB and 10 dB, respectively, below the average power of $C_H$ users (similar results are obtain for MC-CDMA system). From these figures, we can conclude that the iterative procedure allows good performance improvements, especially for $C_L$ users, since the higher BERs associated to $C_L$ users preclude an efficient interference cancelation when detecting $C_H$ users. However, this effect is less significant when the difference between the users' average power increases (see Fig. 3.22).

# 4

# Receivers for the Uplink Transmissions of DS-CDMA and MC-CDMA Systems

This chapter considers the uplink transmission of DS-CDMA and MC-CDMA systems employing CP-assisted block transmission techniques. Frequency-domain MUD receivers for both systems are studied, combining IB-DFE principles and MAI cancelation techniques.

The publications resulting from the work presented in this chapter are [55–59].

This chapter is organized as follows: Section 4.1 gives a brief introductory overview on MUD receivers for DS-CDMA and MC-CDMA systems. Section 4.2 presents the study for the uplink receivers of DS-CDMA system which includes the system characterization, the frequency-domain receivers structure combining turbo equalization with MUD, as well as an extension for MIMO systems and a discussion on system implementation and signal processing complexity issues. Finally, a set of performance results and the corresponding discussion ends this section. Section 4.3 is devoted to the study of the uplink receivers of MC-CDMA system, also including the receivers structure, a brief complexity analysis and a set of performance results.

## 4.1 Introduction

As described in Chapter 3, the receivers for CP-assisted DS-CDMA and MC-CDMA are particularly simple at the downlink: since all spreading codes are affected by the same multipath channel, the receivers can be based on a linear FDE, operating at the chip level, followed by the despreading procedure [37, 43, 60]. The performance can be further improved if the linear FDE is replaced by an IB-DFE, especially for fully loaded scenarios and/or in the presence of strong interference levels [51]. The receiver design for the uplink transmission is more challenging, due to the fact that the signals associated to different users are affected by different propagation channels. Since the orthogonality between spreading codes associated to different users is lost, severe interference levels can be expected, especially for fully loaded systems and/or when different powers are assigned to different spreading codes, leading to significant performance degradation.

To improve the performance several turbo-MUD receivers were proposed for DS-CDMA systems, as well as MC-CDMA. In [47], a successful approach for coded CDMA systems, based on the "turbo principle", is a low-complexity PIC MUD using the soft feedback from the output of the turbo decoder to improve the PIC's performance. A similar approach, but with an MIMO scheme employing space-time turbo codes is proposed in [61]. In [46], the use of soft information for interference cancelation is also exploited, but in MC-CDMA systems. By taking advantage of the implicit diversity effect in the transmitted signals inherent to the spreading procedure in CDMA systems, frequency-domain linear MUD receivers can be defined for both DS-CDMA and MC-CDMA schemes as shown in Chapter 2. As with the downlink case described in Chapter 3, to avoid noise enhancement effects, the receiver is usually optimized in the MMSE sense. This means that it is not able to fully orthogonalize the different users (in fact, it is not even able to eliminate completely the ISI in single-user scenarios). Therefore, significant residual interference levels can be expected, especially for fully loaded scenarios and/or in the presence of strong interference levels. To overcome this problem, in this chapter we consider the use of interference cancelation techniques efficiently combined with frequency-domain MUD receivers [55] for the uplink transmissions in DS-CDMA and MC-CDMA systems. Although this leads to suboptimum receivers, they can have good performance and the implementation complexity is much lower than the optimum receivers (whose complexity grows exponentially with the number of spreading codes [62–64]), even for severely time-dispersive channels. We consider both SIC and PIC structures.

## 4.2 Receivers for the Uplink Transmissions of a DS-CDMA System

### 4.2.1 System Characterization

In this section we consider the uplink transmission in DS-CDMA systems employing CP-assisted block transmission techniques, as depicted in Fig. 4.1.

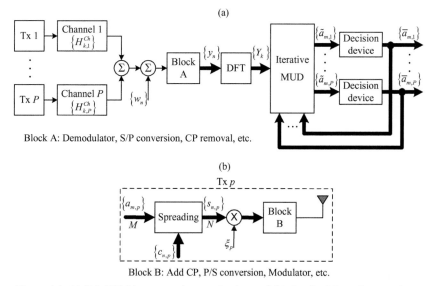

Block A: Demodulator, S/P conversion, CP removal, etc.

Block B: Add CP, P/S conversion, Modulator, etc.

Figure 4.1 (a) DS-CDMA system characterization and (b) detail of the *p*th transmitter.

We have a synchronous system[1] with $P$ users, where the blocks transmitted by each user have the same dimensions. For the sake of simplicity, it is assumed that all users have the same spreading factor $K$ and the same data rate (these techniques can be easily extended to multicode schemes [7]).

As was described in Section 2.3, and briefly repeated here for the sake of easy readability, the size-$M$ data block to be transmitted by the $p$th user is $\{a_{m,p}; m = 0, 1, \ldots, M - 1\}$, with $a_{m,p}$ selected from a given constellation. The corresponding block of chips to be transmitted is $\{s_{n,p}; n = 0, 1, \ldots, N - 1\}$, where $N = MK$ and

$$s_{n,p} = c_{n,p} a_{\lfloor n/K \rfloor, p}. \tag{4.1}$$

It is assumed that the spreading sequences $\{c_{n,p}; n = 0, 1, \ldots, N - 1\}$ are periodic, with period $K$ (i.e., $c_{n+K,p} = c_{n,p}$).

The received signal at the BS is sampled at the chip rate (the generalization for multiple samples per chip is straightforward) and the CP is removed, leading to the time-domain block $\{y_n; n = 0, 1, \ldots, N - 1\}$. Having in mind (2.95)–(2.98), the corresponding frequency-domain block is

---

[1] This means that there is a suitable "time-advance" mechanism allowing perfect synchronization in time at the receiver [58, 65]. However, in practice just a coarse synchronization is required since some time misalignments can be absorbed by the CP.

$\{Y_k; k = 0, 1, \ldots, N - 1\}$, where

$$Y_k = \sum_{p=1}^{P} A_{k \bmod M, p} H_{k,p} + N_k , \qquad (4.2)$$

with

$$H_{k,p} = \frac{1}{K} \xi_p H_{k,p}^{Ch} C'_{k,p} \qquad (4.3)$$

denoting the "overall" channel frequency response for the $p$th user and the $k$th frequency. Using matrix notation, (4.2) is equivalent to

$$\mathbf{Y}(k) = \mathbf{H}^T(k)\mathbf{A}(k) + \mathbf{N}(k) , \qquad (4.4)$$

with matrices $\mathbf{Y}(k)$, $\mathbf{A}(k)$, $\mathbf{N}(k)$ and $\mathbf{H}(k)$ given by (2.66)–(2.70), respectively.

### 4.2.2 Iterative Receiver with Interference Cancelation

In Chapter 2 it was shown that by taking advantage of the spectral correlations that result from the cyclostationarity of the signal transmitted by each MT [45], we can define a frequency-domain linear MUD receiver. To improve the performance, in this section we consider the use of the IB-DFE concept to define an iterative frequency-domain MUD receiver for DS-CDMA system.

**Single-User Scenario**

Let us consider an SU scenario ($p = P = 1$), but with an IB-DFE (with "hard decisions" in the feedback loop) replacing the linear FDE, as shown is Fig. 4.2. In this case, for a given iteration $i$, the frequency-domain samples at the detector output are given by (for the sake of notation simplicity, we will drop the iteration index $i$ from this point forward)

$$\tilde{A}_{k,1} = \mathbf{F}_1^T(k)\mathbf{Y}(k) - B_{k,1}\hat{A}_{k,1} \qquad (4.5)$$

where $\{B_{k,1}; k = 0, 1, \ldots, M - 1\}$ denote the feedback coefficient and $\{\hat{A}_{k,1}; k = 0, 1, \ldots, M-1\}$ the DFT of $\{\hat{a}_{m,1}; m = 0, 1, \ldots, M-1\}$ (i.e., the hard decisions of the samples $\{\tilde{a}_{m,1}; m = 0, 1, \ldots, M - 1\}$ from the previous iteration). The length-$K$ feedforward coefficients vector

$$\mathbf{F}_1(k) = \begin{bmatrix} F_{k,1} \\ F_{k+M,1} \\ \vdots \\ F_{k+(K-1)M,1} \end{bmatrix}, \qquad (4.6)$$

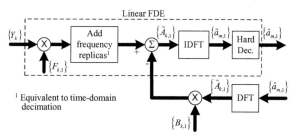

Figure 4.2 Detection of user 1 in a SU scenario with an IB-DFE receiver.

and the feedback coefficient $B_{k,1}$ are chosen so as to maximize the overall SNR in the samples $\{\tilde{a}_{m,1}; m = 0, 1, \ldots, M-1\}$. Since this can be regarded as an IB-DFE receiver with $K$-order space diversity, it can easily be shown that the optimum feedforward and feedback coefficients are given by (see (2.45) and (2.46))

$$\mathbf{F}_1(k) = \left[\left(1 - \rho_1^2\right)\mathbf{H}_1^H(k)\mathbf{H}_1(k) + \beta\mathbf{I}_K\right]^{-1}\mathbf{H}_1^H(k)\gamma_1^{-1} \qquad (4.7)$$

and

$$B_{k,1} = \rho_1\left(\mathbf{H}_1^T(k)\mathbf{F}_1(k) - \gamma_1\right), \qquad (4.8)$$

respectively, where the correlation coefficient $\rho_1$ can be computed as described in Section 3.1.3. Since for the first iteration there is no information about $a_{m,1}$, the correlation coefficient and the feedback coefficient are zero, leading to

$$\mathbf{F}_1(k) = \left[\mathbf{H}_1^H(k)\mathbf{H}_1(k) + \beta\mathbf{I}_K\right]^{-1}\mathbf{H}_1^H(k)\gamma_1^{-1}, \qquad (4.9)$$

which corresponds to the linear FDE receiver, with a despreading operation implicit in the combination of the $K$ "replicas" of $A_{k,1}$ (associated to the time-domain decimation procedure), discussed in Chapter 2. For $K = 1$ (i.e., without spreading), this reduces to the conventional FDE receiver for CP-assisted SC modulations.

## Multiuser Scenario: PIC/SIC Receiver Structures

Let us consider now an iterative frequency-domain MUD receiver that combines IB-DFE principles with interference cancelation. Each iteration consists of $P$ detection stages, one for each user. When detecting a given user, the interference from previously detected users is canceled, as well as the residual ISI associated to that user. These interference and residual ISI cancelations take into account the reliability of each of the previously detected users.

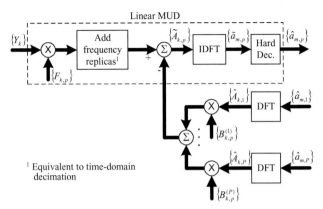

Figure 4.3 Detection of the $p$th user with an IB-DFE receiver.

For a given iteration, the detection of the $p$th user employs the structure depicted in Fig. 4.3, where we have a feedforward filter $\{F_{k,p}; k = 0, ..., N - 1\}$, followed by a decimation procedure and $P$ feedback filters $\{B_{k,p}^{(p')}; p' = 1, ..., P\}$ (one for each user). The feedforward filter is designed to minimize both ISI and interference that cannot be canceled by the feedback filters, due to decision errors in the previous detection steps. After an IDFT operation, the corresponding time-domain outputs $\{\tilde{a}_{m,p}; m = 0, ..., M - 1\}$ are passed through a hard-decision device so as to provide an estimate $\{\hat{a}_{m,p}; m = 0, ..., M - 1\}$ of the data block transmitted by the $p$th user. For the case where we do not have any information about the users' data blocks, the receiver reduces to the linear frequency-domain MUD.

It is assumed that the users are ordered in descending order of their power. We can consider either a SIC or a PIC MUD receiver[2] as shown in Figs. 4.4 and 4.5, respectively. For the SIC receiver, we cancel the interference from all users using the most updated version of it, as well as the residual ISI for the user that is being detected. For the PIC receiver, we cancel the interference from all users, as well as the residual ISI for the user that is being detected, employing the users' estimates from the previous iteration. In general, the achievable performance is similar for both schemes, although the convergence is faster for the SIC receiver, provided that we detect first the MTs for which the power at the BS is higher. The main advantage of the PIC

---

[2] It should be noted that our SIC and PIC receivers are iterative in the sense that each user is estimated several times, while some literature define a SIC receiver where each user is estimated only once.

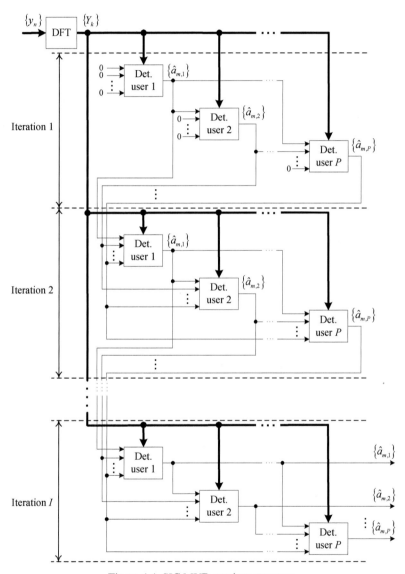

Figure 4.4 SIC MUD receiver structure.

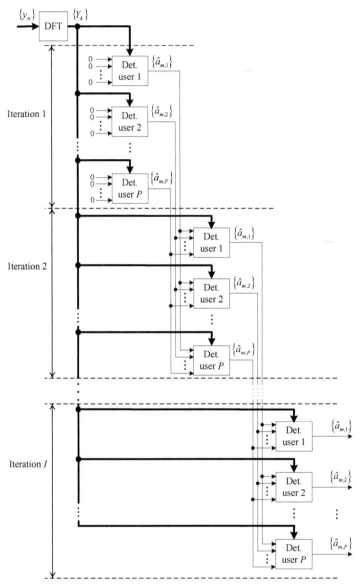

Figure 4.5 PIC MUD receiver structure.

structure is the possibility of a parallel implementation, with the simultaneous detection of all users at each iteration.

### Derivation of the Receiver Parameters

Let us consider first a SIC receiver. For each iteration, the frequency-domain samples associated with the $p$th user at the detector output are given by

$$\tilde{A}_{k,p} = \mathbf{F}_p^T(k)\mathbf{Y}(k) - \mathbf{B}_p^T(k)\hat{\mathbf{A}}(k), \tag{4.10}$$

where the length-$K$ column vector with the feedforward coefficients associated to the $p$th user is

$$\mathbf{F}_p(k) = \begin{bmatrix} F_{k,p} \\ F_{k+M,p} \\ \vdots \\ F_{k+(K-1)M,p} \end{bmatrix} \tag{4.11}$$

and the feedback coefficients length-$P$ column vector

$$\mathbf{B}_p(k) = \begin{bmatrix} B_{k,p}^{(1)} \\ \vdots \\ B_{k,p}^{(P)} \end{bmatrix}. \tag{4.12}$$

The length-$P$ column vector $\hat{\mathbf{A}}(k)$ is defined as

$$\hat{\mathbf{A}}(k) = \begin{bmatrix} \hat{A}_{k,1} \\ \vdots \\ \hat{A}_{k,P} \end{bmatrix} \tag{4.13}$$

and the block $\{\hat{A}_{k,p'}; k = 0, 1, \ldots, M - 1\}$ is the DFT of the block $\{\hat{a}_{m,p'}; m = 0, 1, \ldots, M - 1\}$, with $\{\hat{a}_{m,p'}; m = 0, 1, \ldots, M - 1\}$ denoting the latest estimates for the $p'$th user transmitted symbols (i.e., the hard decisions of the samples $\{\tilde{a}_{m,p'}; m = 0, 1, \ldots, M - 1\}$ from the previous iteration).

For a PIC receiver, the frequency-domain samples associated to all users at the detector output are given by

$$\tilde{\mathbf{A}}(k) = \mathbf{F}^T(k)\mathbf{Y}(k) - \mathbf{B}^T(k)\hat{\mathbf{A}}(k), \tag{4.14}$$

with the feedforward $K \times P$ matrix

$$\mathbf{F}(k) = [\mathbf{F}_1(k) \ \ldots \ \mathbf{F}_P(k)] \tag{4.15}$$

and the feedback $P \times P$ matrix

$$\mathbf{B}(k) = [\mathbf{B}_1(k) \ \ldots \ \mathbf{B}_P(k)].$$  (4.16)

For the $i$th iteration of a SIC receiver, $\hat{a}_{m,p'}$ is associated with the $i$th iteration for $p' < p$ and with the $(i - 1)$th iteration for $p' \geq p$ (in the first iteration, we do not have any information for $p' \geq p$ and $\hat{a}_{m,p'} = 0$) (see Fig. 4.4); for a PIC receiver, $\hat{a}_{m,p'}$ is always associated with the previous iteration (for the first iteration $\hat{a}_{m,p'} = 0$) (see Fig. 4.5).

Due to decision errors, we have $\hat{a}_{m,p} \neq a_{m,p}$ for some symbols. Consequently, $\hat{A}_{k,p} \neq A_{k,p}$. To simplify the computation of the receiver coefficients, it is assumed that

$$\hat{A}_{k,p} = \rho_p A_{k,p} + \Delta_{k,p},$$  (4.17)

where $E[\Delta_{k,p}] \approx 0$ and $E[\Delta_{k,p} A_{k',p}] \approx 0$, regardless of $k$ and $k'$ (a similar assumption was made in Chapter 3 (see (3.6) and (3.8))). Moreover,

$$E\left[\left|\Delta_{k,p}\right|^2\right] = \left(1 - \rho_p^2\right) E\left[\left|A_{k,p}\right|^2\right].$$  (4.18)

In matrix notation, (4.17) takes the form[3]

$$\hat{\mathbf{A}}(k) = \mathbf{P}\mathbf{A}(k) + \mathbf{\Delta}(k)$$  (4.19)

with the length-$P$ column vector

$$\mathbf{\Delta}(k) = \begin{bmatrix} \Delta_{k,1} \\ \vdots \\ \Delta_{k,P} \end{bmatrix}$$  (4.20)

and

$$\mathbf{P} = \mathrm{diag}(\rho_1, \ldots, \rho_P).$$  (4.21)

By combining (4.4), (4.10) and (4.19), we obtain, after some straightforward manipulation,

$$\tilde{A}_{k,p} = \underbrace{\mathbf{\Gamma}_p^T(k)\mathbf{A}(k)}_{\text{Useful signal}} + \underbrace{\left(\mathbf{F}_p^T(k) \ \mathbf{H}^T(k) - \mathbf{B}_p^T(k) \ \mathbf{P} - \mathbf{\Gamma}_p^T(k)\right) \mathbf{A}(k)}_{\text{Residual Interference (ISI + MAI)}}$$

$$- \underbrace{\mathbf{B}_p^T(k) \ \mathbf{\Delta}(k)}_{\substack{\text{``Noise'' due to} \\ \text{feedback errors}}} + \underbrace{\mathbf{F}_p^T(k) \ \mathbf{N}(k)}_{\text{Channel noise}}.$$  (4.22)

---

[3]  $\mathbf{P}$ (capital $\rho$) should not be confused with $P$ (number of users).

For a SIC receiver, the forward and backward vectors, $\mathbf{F}_p(k)$ and $\mathbf{B}_p(k)$, respectively, are chosen so as to maximize the $\text{SINR}_p$ for the $p$th user, at a particular iteration, defined as

$$\text{SINR}_p = \frac{|\gamma_p|^2 E\left[|A_{k,p}|^2\right]}{E\left[|\Theta_{k,p}|^2\right]} \tag{4.23}$$

where

$$\Theta_{k,p} = \tilde{A}_{k,p} - \mathbf{\Gamma}_p^T(k)\mathbf{A}(k) \tag{4.24}$$

denotes the "overall" error for the $p$th user that includes both the channel noise and the residual interference in the frequency-domain. The maximization of (4.23) is equivalent to the minimization of

$$E\left[|\Theta_{k,p}|^2\right] = E\left[\left|\left(\mathbf{F}_p^T(k)\mathbf{H}^T(k) - \mathbf{B}_p^T(k)\mathbf{P} - \mathbf{\Gamma}_p^T(k)\right)\mathbf{A}(k)\right|^2\right]$$
$$+ E\left[\left|\mathbf{B}_p^T(k)\mathbf{\Delta}(k)\right|^2\right] + E\left[\left|\mathbf{F}_p^T(k)\mathbf{N}(k)\right|^2\right] \tag{4.25}$$

By employing the Lagrange multipliers' method, the optimum receiver vectors $\mathbf{F}_p(k)$ and $\mathbf{B}_p(k)$ can be obtained. For this purpose, we can define the Lagrange function

$$J = E\left[|\Theta_{k,p}|^2\right] + \lambda_p(\gamma_p - 1), \tag{4.26}$$

where $\{\lambda_p,\ p = 1, ..., P\}$, are the Lagrange multipliers, and assume that the optimization is carried out under $\gamma_p = 1$. The optimum feedforward vectors are obtained by solving the following set of equations:

- $\nabla_{\mathbf{F}_p(k)} J = 0 \Leftrightarrow \mathbf{H}^H(k)\mathbf{R}_A\mathbf{H}(k)\mathbf{F}_p(k) - \mathbf{H}^H(k)\mathbf{R}_A\mathbf{P}\mathbf{B}_p(k)$

$$-\mathbf{H}^H(k)\mathbf{R}_A\mathbf{\Gamma}_p(k) + \mathbf{R}_N\mathbf{F}_p(k) + \frac{\lambda_p}{M}\mathbf{H}^H(k) = 0$$
$$\Leftrightarrow \mathbf{H}^H(k)\mathbf{H}(k)\mathbf{F}_p(k) - \mathbf{H}^H(k)\mathbf{P}\mathbf{B}_p(k) - \mathbf{H}^H(k)\mathbf{\Gamma}_p(k)$$
$$+\beta\mathbf{F}_p(k) + \frac{\lambda_p}{2\sigma_A^2 M}\mathbf{H}_p^H(k) = 0, \tag{4.27}$$

with $\mathbf{R}_A = E[\mathbf{A}^*(k)\mathbf{A}^T(k)] = 2\sigma_A^2\mathbf{I}_P$ and $\mathbf{R}_N = E[\mathbf{N}^*(k)\mathbf{N}^T(k)] = 2\sigma_N^2\mathbf{I}_K$;

- $\nabla_{\mathbf{B}_p(k)} J = 0 \Leftrightarrow (\mathbf{P}^2\mathbf{R}_A + \mathbf{R}_\Delta)\mathbf{B}_p(k) = \mathbf{P}\mathbf{R}_A\mathbf{H}(k)\mathbf{F}_p(k)$ \tag{4.28}
$$-\mathbf{P}\mathbf{R}_A\mathbf{\Gamma}_p(k),$$

with $\mathbf{R}_\Delta = E[\boldsymbol{\Delta}^*(k)\boldsymbol{\Delta}^T(k)] = 2\sigma_A^2 \operatorname{diag}(1-\rho_1^2, \ldots, 1-\rho_P^2) = 2\sigma_A^2(\mathbf{I}_P - \mathbf{P}^2)$;

- $\nabla_{\lambda_p} J = 0 \Leftrightarrow \gamma_p = 1.$ \hfill (4.29)

As expected, (4.29) is the condition under which the optimization is carried out.

Rewriting (4.28) by noting that

$$\mathbf{P}^2\mathbf{R}_A + \mathbf{R}_\Delta = 2\sigma_A^2\mathbf{P}^2 + 2\sigma_A^2(\mathbf{I}_P - \mathbf{P}^2) = 2\sigma_A\mathbf{I}_P = \mathbf{R}_A, \tag{4.30}$$

the optimum feedback vector $\mathbf{B}_p(k)$ is given by

$$\mathbf{B}_p(k) = \mathbf{P}\left(\mathbf{H}(k)\mathbf{F}_p(k) - \boldsymbol{\Gamma}_p(k)\right). \tag{4.31}$$

By replacing (4.31) in (4.27), the optimum feedforward vector $\mathbf{F}_p(k)$ can be written as

$$\mathbf{F}_p(k) = \left[\mathbf{H}^H(k)(\mathbf{I}_P - \mathbf{P}^2)\mathbf{H}(k) + \beta\mathbf{I}_K\right]^{-1}\mathbf{H}^H(k)\mathbf{Q}_p, \tag{4.32}$$

where the constant normalization length-$P$ column vector

$$\mathbf{Q}_p = \begin{bmatrix} 0 \\ \vdots \\ Q_p \\ \vdots \\ 0 \end{bmatrix} = \left(\mathbf{I}_P - \mathbf{P}^2\right)\boldsymbol{\Gamma}_p(k) - \frac{1}{2\sigma_A^2 M}\boldsymbol{\Lambda}_p, \tag{4.33}$$

with

$$\boldsymbol{\Lambda}_p = \begin{bmatrix} 0 \\ \vdots \\ \lambda_p \\ \vdots \\ 0 \end{bmatrix}, \tag{4.34}$$

i.e., $\boldsymbol{\Lambda}_p$ is a length-$P$ column vector that has the Lagrange's multiplier associated to the $p$th user in the $p$th position. This vector ensures that $\gamma_p = 1$ or equivalently

$$\boldsymbol{\Gamma}_p(k) = \begin{bmatrix} 0 \\ \vdots \\ 1 \\ \vdots \\ 0 \end{bmatrix}. \tag{4.35}$$

Similarly, for a PIC receiver, the optimum forward and backward matrices, $\mathbf{F}(k)$ and $\mathbf{B}(k)$, respectively, can also be obtained by employing the Lagrange multipliers' method. In this case,

$$\mathbf{F}(k) = \left[\mathbf{H}^H(k)(\mathbf{I}_P - \mathbf{P}^2)\mathbf{H}(k) + \beta\mathbf{I}_K\right]^{-1}\mathbf{H}^H(k)\mathbf{Q}, \qquad (4.36)$$

and

$$\mathbf{B}(k) = \mathbf{P}\left(\mathbf{H}(k)\mathbf{F}(k) - \mathbf{\Gamma}(k)\right), \qquad (4.37)$$

where the constant normalization $P \times P$ diagonal matrix

$$\mathbf{Q} = \left[\mathbf{Q}_1 \ldots \mathbf{Q}_P\right] = \mathrm{diag}(Q_1 \ldots Q_P) = \left(\mathbf{I}_P - \mathbf{P}^2\right)\mathbf{\Gamma}(k) - \frac{1}{2\sigma_A^2 M}\mathbf{\Lambda}. \quad (4.38)$$

It should be noted that the feedforward and feedback coefficients obtained this way are closely related to the optimum coefficients for the iterative layered space-time receivers proposed in [66]. This results from the fact that frequency-domain receivers with ISI and interference cancelation are considered in both cases (in our case we have MAI and in [66] there is "inter-layer" interference); moreover, in both cases, the user/layer separation is based on multiple replicas of each frequency-domain sample (in our case this multiplicity is inherent to the transmitted signals, while, in [66], it results from employing multiple antennas at the receiver).

If we do not have data estimates for the different users $\mathbf{P} = 0$ and the feedback coefficients are zero. Therefore, (4.36) reduces to (2.89), i.e.,

$$\mathbf{F}(k) = \left[\mathbf{H}^H(k)\mathbf{H}(k) + \beta\mathbf{I}_K\right]^{-1}\mathbf{H}^H(k)\mathbf{Q} \qquad (4.39)$$

corresponding to the linear MUD described in Chapter 2.

In Appendix D, it is shown that the optimum feedforward vector given by (4.32) can also be written as

$$\mathbf{F}_p(k) = \mathbf{H}^H(k)\mathbf{V}(k)\mathbf{Q}_p \qquad (4.40)$$

for a SIC receiver, where the $P \times P$ matrix $\mathbf{V}(k)$ is given by

$$\mathbf{V}(k) = \left[(\mathbf{I}_P - \mathbf{P}^2)\mathbf{H}(k)\mathbf{H}^H(k) + \beta\mathbf{I}_P\right]^{-1}. \qquad (4.41)$$

Following the same reasoning, (4.36) can also be written as

$$\mathbf{F}(k) = \mathbf{H}^H(k)\mathbf{V}(k)\mathbf{Q}, \qquad (4.42)$$

for a PIC receiver.

The computation of the feedforward coefficients from (4.40)–(4.42) is simpler than the direct computation, from (4.32) and (4.36), especially when $P < K$.

## Soft Decisions

As it was described in Section 3.2, we can improve the performance by using "symbol averages" instead of "blockwise averages" in the feedback loop of the IB-DFE. Similarly to (3.38), (4.10) and (4.14) can be written as

$$\tilde{A}_{k,p} = \mathbf{F}_p^T(k)\mathbf{Y}(k) - \mathbf{B}_p^T(k)\overline{\mathbf{A}}(k) \tag{4.43}$$

and

$$\tilde{\mathbf{A}}(k) = \mathbf{F}^T(k)\mathbf{Y}(k) - \mathbf{B}^T(k)\overline{\mathbf{A}}(k), \tag{4.44}$$

respectively. $\overline{\mathbf{A}}(k)$ is the column vector with the average frequency-domain samples $\{\overline{A}_{k,p} = \rho_p\hat{A}_{k,p}; k = 0, 1, \ldots, M - 1\}$ (i.e., the DFT of the block $\{\overline{a}_{m,p} = \rho_p\hat{a}_{m,p}; m = 0, 1, \ldots, M - 1\}$).

As with (3.46) (see also Appendix C), the mean value of $a_{m,p}$ is given by

$$\overline{a}_{m,p} = \tanh\left(\frac{L_{m,p}^I}{2}\right) + j\tanh\left(\frac{L_{m,p}^Q}{2}\right), \tag{4.45}$$

with the LLRs of the "in-phase" and the "quadrature" bits, associated to $a_{m,p}^I$ and $a_{m,p}^Q$, respectively, given by

$$L_{m,p}^I = \frac{2}{\sigma_p^2}\tilde{a}_{m,p}^I \tag{4.46}$$

and

$$L_{m,p}^Q = \frac{2}{\sigma_p^2}\tilde{a}_{m,p}^Q, \tag{4.47}$$

where

$$\sigma_p^2 = \frac{1}{2}E\left[|a_{m,p} - \tilde{a}_{m,p}|^2\right] \approx \frac{1}{2M}\sum_{m=0}^{M-1}|\hat{a}_{m,p} - \tilde{a}_{m,p}|^2. \tag{4.48}$$

The feedforward coefficients are still obtained from (4.32) (or (4.40) and (4.41)), but with the blockwise reliability given by (3.49). The receiver with "blockwise reliabilities" will be denoted in the following as IMUD-HD (Iterative MUD with Hard Decisions) and the receiver with "symbol reliabilities" as IMUD-SD (Iterative MUD with Soft Decisions).

**Use of Channel Decoder Outputs in the Feedback Loop**

Similarly to the turbo equalizers for the downlink transmissions treated in Section 3.2.1, it is also possible to define a receiver with joint equalization and multiuser detection that employs the channel decoder outputs (instead of the "soft decisions" from the IMUD) in the feedback loop. The receiver structure, that will be denoted as Turbo MUD-SD (Turbo MUD with Soft Decisions), is similar to the IMUD-SD, but with a SISO channel decoder employed in the feedback loop. The input of the SISO block are LLRs of the "coded bits" at the multiuser detector, given by (4.46) and (4.47). Once again, the feedforward coefficients are obtained from (4.32) (or (4.40) and (4.41)), with the blockwise reliability given by (3.49).

As an alternative, we could employ a conventional decoder, unable to provide the LLRs of the "coded bits" but allowing a simpler implementation (e.g., a Viterbi decoder in the case of convolutional codes). Since we are not able to obtain the reliabilities of each bit, we can assume that $\rho_p = \rho_{m,p}^I = \rho_{m,p}^Q = 1$, which is a simple, but suboptimal, solution. This receiver will be denoted Turbo MUD-HD (Turbo MUD with Hard Decisions).

**Signal Processing Complexity**

The implementation complexity of the proposed receivers can be measured in terms of the number and size of the DFT/IDFT operations and the computation charge required for the calculation of the feedforward and feedback coefficients. In the case of the IMUD receivers, both SIC and PIC structures require a size-$N$ DFT operation, a pair of size-$M$ DFT/IDFT operations (except for the first iteration where only a size-$M$ IDFT is required) and a decision operation for the detection of each user, at each iteration. As for the computation of the feedforward coefficients, we need to invert the $P \times P$ matrix (4.41), for each iteration and each user. Naturally, for slowly-varying channels, this operation is not required for all blocks. In the case of the Turbo MUD-SD receiver, the SISO channel decoding needs to be implemented in the detection process of each user, with the soft-output Viterbi algorithm instead of a conventional Viterbi algorithm. This can be the most complex part of the Turbo MUD-SD receiver. Nevertheless, it should be pointed out that the implementation charge is concentrated at the BS, where increased power consumption and cost are not so critical.

It should also be pointed out that whenever $\rho_p \approx 1$ for the $p$th user at a given iteration, we already have reliable decisions for that user and its interference can be removed almost entirely when detecting the other users. This means that this user can be excluded from the detection process in the next iterations, decreasing the implementation complexity with only negligible performance degradation.

### 4.2.3 Extension for MIMO Systems

This section presents an extension of the iterative frequency-domain receivers for DS-CDMA systems studied in the previous section, to multi-antenna scenarios. We consider spacial multiplexing techniques [66,67] which allow significant increase in the system's spectral efficiency and require multiple antennas at both the transmitter and the receiver.

**System Characterization**

The system considered in this section is depicted in Fig. 4.6. The BS has $L_R$ receive antennas and the $p$th MT has $L_T^{(p)}$ transmit antennas, each one transmitting a different stream of data symbols. It is assumed that the received blocks associated to each MT are synchronized in time.

The size-$M$ data block to be transmitted by the $l$th antenna of the $p$th MT is $\{a_{m,l}^{(p)}; m = 0, 1, \ldots, M-1\}$, with $a_{m,l}^{(p)}$ selected from a given constellation. The corresponding chip block to be transmitted is $\{s_{n,l}^{(p)}; n = 0, 1, \ldots, N-1\}$, where $N = MK$ and

$$s_{n,l}^{(p)} = c_{n,l}^{(p)} a_{\lfloor n/K \rfloor, l}^{(p)}, \tag{4.49}$$

with $c_{n,l}^{(p)}$ denoting the spreading symbols[4] (again, it is assumed that the spreading sequence is periodic, with period $K$, i.e., $c_{n+K,l}^{(p)} = c_{n,l}^{(p)}$).

The received signal associated to the $r$th antenna of the BS is sampled at the chip rate and the CP is removed, leading to the time-domain block $\{y_n^{(r)}; n = 0, 1, \ldots, N-1\}$. The corresponding frequency-domain block is $\{Y_k^{(r)}; k = 0, 1, \ldots, N-1\}$, where

$$Y_k^{(r)} = \sum_{p=1}^{P} \sum_{l=1}^{L_T^{(p)}} S_{k,l}^{(p)} \xi_l^{(p)} H_{k,l}^{Ch,p,r} + N_k^{(r)} \tag{4.50}$$

with $H_{k,l}^{Ch,p,r}$ denoting the channel frequency response between the $l$th transmit antenna of the $p$th MT and the $r$th receive antenna of the BS, at the $k$th frequency (without loss of generality, it is assumed that $E[|H_{k,l}^{Ch,p,r}|^2] = 1$).

---

[4] It will be shown in the following that the different transmit antennas associated to a given MT can have the same spreading code or not.

(a)

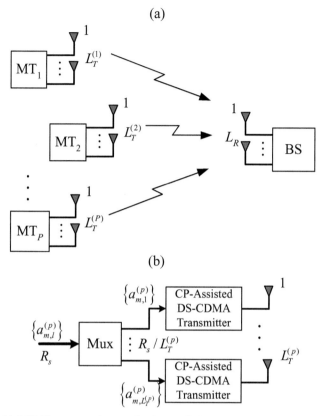

(b)

Figure 4.6 (a) MIMO system characterization and (b) detail of the $p$th MT, with spatial multiplexing of degree $L_T^{(p)}$ and data rate $R_S$.

$N_k^{(r)}$ is the channel noise at the $r$th receive antenna, for the $k$th frequency and $\xi_l^{(p)}$ is a scale factor that accounts for the combined effect of the propagation losses and the power assigned to the $l$th antenna of the $p$th MT. The frequency-domain block $\{S_{k,l}^{(p)}; k = 0, 1, \ldots, N - 1\}$ is the DFT of the chip block transmitted by the $l$th antenna of the $p$th MT, $\{s_{n,l}^{(p)}; n = 0, 1, \ldots, N - 1\}$.

Extending the analysis made for the single-layer case to the MIMO case, concerning the $k$-order diversity effect related to the cyclostationary nature of the transmitted signals described in Section 2.3 (see (2.95)–(2.98) and Fig. 2.31), it can be easily shown that

$$S_{k,l}^{(p)} = A_{k,l}^{'(p)} C_{k,l}^{'(p)} \tag{4.51}$$

where $\{A_{k,l}^{'(p)}; k = 0, 1, \ldots, N - 1\}$ denotes the DFT of the block $\{a_{n,l}^{'(p)}; n = 0, 1, \ldots, N - 1\}$, with

$$a_{n,l}^{'(p)} = \begin{cases} a_{m,l}^{(p)}, & n = mK \\ 0, & \text{otherwise} \end{cases} \tag{4.52}$$

and $\{C_{k,l}^{'(p)}; k = 0, 1, \ldots, N - 1\}$ denotes the DFT of the block $\{c_{n,l}^{'(p)}; n = 0, 1, \ldots, N - 1\}$, with

$$c_{n,l}^{'(p)} = \begin{cases} c_{n,l}^{(p)}, & 0 \leq n < K - 1 \\ 0, & \text{otherwise} \end{cases} . \tag{4.53}$$

This means that

$$A_{k,l}^{'(p)} = \frac{1}{K} A_{k \bmod M,l}^{(p)} , \tag{4.54}$$

$k = 0, 1, \ldots, N - 1$, with $\{A_{k,l}^{(p)}; k = 0, 1, \ldots, M - 1\}$ denoting the DFT of the block $\{a_{m,l}^{(p)}; m = 0, 1, \ldots, M - 1\}$. Therefore,

$$Y_k^{(r)} = \sum_{p=1}^{P} \sum_{l=1}^{L_T^{(p)}} A_{k \bmod M,l}^{(p)} H_{k,l}^{(p,r)} + N_k^{(r)}, \tag{4.55}$$

with

$$H_{k,l}^{(p,r)} = \frac{1}{K} \xi_l^{(p)} H_{k,l}^{Ch,p,r} C_{k,l}^{'(p)} \tag{4.56}$$

denoting the equivalent channel frequency response between the $l$th transmit antenna of the $p$th MT and the $r$th receive antenna of the BS, for the $k$th frequency. In matrix notation (4.55) is equivalent to

$$\mathbf{Y}_k = \mathbf{H}_k^T \mathbf{A}_k + \mathbf{N}_k, \tag{4.57}$$

where the length-$KL_R$ column vectors $\mathbf{Y}_k$ and $\mathbf{N}_k$ are, respectively, given by

$$\mathbf{Y}_k = \begin{bmatrix} Y_k^{(1)} \\ \vdots \\ Y_k^{(L_R)} \\ \vdots \\ Y_{k+(K-1)M}^{(1)} \\ \vdots \\ Y_{k+(K-1)M}^{(L_R)} \end{bmatrix}, \tag{4.58}$$

$$\mathbf{N}_k = \begin{bmatrix} N_k^{(1)} \\ \vdots \\ N_k^{(L_R)} \\ \vdots \\ N_{k+(K-1)M}^{(1)} \\ \vdots \\ N_{k+(K-1)M}^{(L_R)} \end{bmatrix}. \tag{4.59}$$

Clearly, the column vector

$$\mathbf{A}_k = \begin{bmatrix} A_{k \bmod M,1}^{(1)} \\ \vdots \\ A_{k \bmod M, L_T^{(1)}}^{(1)} \\ \vdots \\ A_{k \bmod M,1}^{(P)} \\ \vdots \\ A_{k \bmod M, L_T^{(P)}}^{(P)} \end{bmatrix} \tag{4.60}$$

has length $N_L = \sum_{p=1}^{P} L_T^{(p)}$, with $N_L$ denoting the total number of transmitted layers. The size-$N_L \times KL_R$ matrix

$$
\begin{aligned}
\mathbf{H}_k &= \begin{bmatrix}
H_{k,1}^{(1,1)} & \cdots & H_{k,1}^{(1,L_R)} & H_{k+(K-1)M,1}^{(1,1)} & \cdots & H_{k+(K-1)M,1}^{(1,L_R)} \\
\vdots & & \vdots & \vdots & & \vdots \\
H_{k,L_T^{(1)}}^{(1,1)} & \cdots & H_{k,L_T^{(1)}}^{(1,L_R)} & H_{k+(K-1)M,L_T^{(1)}}^{(1,1)} & \cdots & H_{k+(K-1)M,L_T^{(1)}}^{(1,L_R)} \\
& \vdots & & & \vdots & \\
H_{k,1}^{(P,1)} & \cdots & H_{k,1}^{(P,L_R)} & H_{k+(K-1)M,1}^{(P,1)} & \cdots & H_{k+(K-1)M,1}^{(P,L_R)} \\
\vdots & & \vdots & \vdots & & \vdots \\
H_{k,L_T^{(P)}}^{(P,1)} & \cdots & H_{k,L_T^{(P)}}^{(P,L_R)} & H_{k+(K-1)M,L_T^{(P)}}^{(P,1)} & \cdots & H_{k+(K-1)M,L_T^{(P)}}^{(P,L_R)}
\end{bmatrix} \\[2ex]
&= \begin{bmatrix}
\mathbf{H}_0^{(1)} & \cdots & \mathbf{H}_{K-1}^{(1)} \\
\vdots & & \vdots \\
\mathbf{H}_0^{(P)} & \cdots & \mathbf{H}_{K-1}^{(P)}
\end{bmatrix},
\end{aligned}
\tag{4.61}
$$

with

$$
\mathbf{H}_q^{(p)} =
\begin{bmatrix}
H_{k+qM,1}^{(p,1)} & \cdots & H_{k+qM,1}^{(p,L_R)} \\
\vdots & & \vdots \\
H_{k+qM,L_T^{(p)}}^{(p,1)} & \cdots & H_{k+qM,L_T^{(p)}}^{(p,L_R)}
\end{bmatrix}.
\tag{4.62}
$$

Since we have $KL_R$ replicas associated to each $A_{k,l}^{(p)}$ we can separate $KL_R$ different transmitted layers at the BS, i.e., we should have

$$
N_L = \sum_{p=1}^{P} L_T^{(p)} \leq KL_R,
\tag{4.63}
$$

for an ideal separation.[5]

**Receiver Structure**

Similarly to the single-layer case, each iteration consists of $N_L$ detection stages, since in this case there is $N_L$ different layers. When detecting a given layer, the interference from the other layers and the residual ISI associated to that layer are canceled taking into account the reliability of each of the previously detected layers.

The detection of the $l$th layer of the $p$th MT at a given iteration employs the structure depicted in Fig. 4.7. In this case, $L_R$ feedforward filters (one for each receive antenna) and $N_L$ feedback filters (one for each layer) are needed. Naturally, both SIC and PIC MUD schemes of Figs. 4.4 and 4.5, respectively, can also be considered but instead of "users detection" we have "layers detection". An estimate of the data block transmitted by the $l$th antenna of the $p$th MT $\{a_{m,l}^{(p)}; m = 0, 1, \ldots, M - 1\}$ is provided by submitting the corresponding time-domain samples through a hard-decision device.[6] For the first iteration, the receiver reduces to a linear frequency-domain MUD with $L_R$-branch space diversity, since no information exists about the layers' data blocks.

---

[5] For an overloaded system, (4.63) does not hold. However, it should be noted that our receiver might still be able to separate the layers in slightly overloaded systems, although with some performance degradation.

[6] Naturally, a soft-decision device can also be used instead of an hard-decision one, following the same approach previously described for the single-layer case in Section 4.2.2.

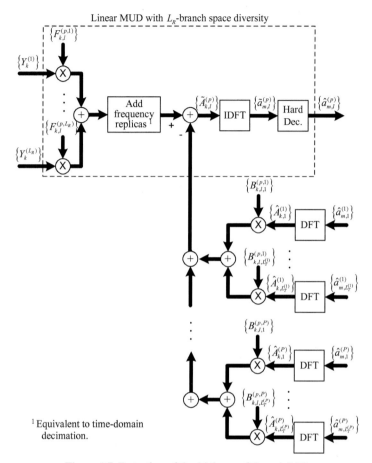

Figure 4.7 Detection of the *l*th layer of the *p*th MT.

For each iteration, the frequency-domain samples associated with the *l*th antenna of the *p*th MT at the detector's output are given by

$$\tilde{A}_{k,l}^{(p)} = \sum_{r=1}^{L_R} \sum_{q=0}^{K-1} F_{k+qM,l}^{(p,r)} Y_{k+qM}^{(r)} - \sum_{p'=1}^{P} \sum_{l'=1}^{L_T^{(p')}} B_{k,l,l'}^{(p,p')} \hat{A}_{k,l'}^{(p')} \qquad (4.64)$$

where $\{F_{k,l}^{(p,r)}; k = 0, 1, \ldots, N-1; r = 1, 2, \ldots, L_R\}$ denote the feed-forward coefficients and $\{B_{k,l,l'}^{(p,p')}; k = 0, 1, \ldots, M-1; p' = 1, 2, \ldots, P; l' = 1, 2, \ldots, L_T^{(p')}\}$ denote the feedback coefficients. The coefficients

$\{B_{k,l,l}^{(p,p)}; k = 0, 1, \ldots, M - 1\}$ are used for residual ISI cancelation and the coefficients $\{B_{k,l,l'}^{(p,p')}; k = 0, 1, \ldots, M - 1\}$ ($l' \neq l$ or $p \neq p'$) are used for interference cancelation. The block $\{\hat{A}_{k,l'}^{(p')}; k = 0, 1, \ldots, M - 1\}$ is the DFT of the block $\{\hat{a}_{m,l'}^{(p')}; m = 0, 1, \ldots, M - 1\}$, where the time-domain samples $\hat{a}_{m,l'}^{(p')}, m = 0, 1, \ldots, M - 1$, are the latest estimates for the transmitted symbols of the $l'$th antenna of the $p'$th MT, i.e., the hard-decisions associated with the block of time-domain samples $\{\tilde{a}_{m,l'}^{(p')}; m = 0, 1, \ldots, M - 1\}$, the IDFT of $\{\tilde{A}_{k,l'}^{(p')}; k = 0, 1, \ldots, M - 1\}$. For the $i$th iteration of a SIC receiver, $\hat{a}_{m,l'}^{(p')}$ is associated with the $i$th iteration for $l' < l$ and with the $(i - 1)$th iteration for $l' \geq l$ (in the first iteration, we do not have any information for $l' \geq l$ and $\hat{a}_{m,l'}^{(p')} = 0$); for a PIC receiver, $\hat{a}_{m,l'}^{(p')}$ is always associated with the previous iteration, with $\hat{a}_{m,l'}^{(p')} = 0$ for the first iteration.

For a SIC receiver, (4.64) can be rewritten in matrix form as

$$\tilde{A}_{k,l}^{(p)} = \mathbf{F}_{k,l}^{(p)^T} \mathbf{Y}_k - \mathbf{B}_{k,l}^{(p)^T} \hat{\mathbf{A}}_k \tag{4.65}$$

and for a PIC receiver as

$$\tilde{\mathbf{A}}_k = \mathbf{F}_k^T \mathbf{Y}_k - \mathbf{B}_k^T \hat{\mathbf{A}}_k, \tag{4.66}$$

where vectors $\hat{\mathbf{A}}_k$ and $\tilde{\mathbf{A}}_k$ are similarly defined as $\mathbf{A}_k$ by (4.60), the size-$KL_R \times N_L$ feedforward coefficients matrix

$$\mathbf{F}_k = \begin{bmatrix} \mathbf{F}_{k,1}^{(1)} \ldots \mathbf{F}_{k,L_T^{(1)}}^{(1)} & \cdots & \mathbf{F}_{k,1}^{(P)} \ldots \mathbf{F}_{k,L_T^{(P)}}^{(P)} \end{bmatrix}$$

$$= \begin{bmatrix} F_{k,1}^{(1,1)} & \cdots & F_{k,L_T^{(1)}}^{(1,1)} & F_{k,1}^{(P,1)} & \cdots & F_{k,L_T^{(P)}}^{(P,1)} \\ \vdots & & \vdots & \cdots & \vdots & & \vdots \\ F_{k,1}^{(1,L_R)} & \cdots & F_{k,L_T^{(1)}}^{(1,L_R)} & F_{k,1}^{(P,L_R)} & \cdots & F_{k,L_T^{(P)}}^{(P,L_R)} \\ \vdots & & \vdots & & \vdots & & \vdots \\ F_{k+(K-1)M,1}^{(1,1)} & \cdots & F_{k+(K-1)M,L_T^{(1)}}^{(1,1)} & F_{k+(K-1)M,1}^{(P,1)} & \cdots & F_{k+(K-1)M,L_T^{(P)}}^{(P,1)} \\ \vdots & & \vdots & \cdots & \vdots & & \vdots \\ F_{k+(K-1)M,1}^{(1,L_R)} & \cdots & F_{k+(K-1)M,L_T^{(1)}}^{(1,L_R)} & F_{k+(K-1)M,1}^{(P,L_R)} & \cdots & F_{k+(K-1)M,L_T^{(P)}}^{(P,L_R)} \end{bmatrix},$$

$$\tag{4.67}$$

and the size-$N_L \times N_L$ feedback coefficients matrix

$$\mathbf{B}_k = \left[ \mathbf{B}_{k,1}^{(1)} \ldots \mathbf{B}_{k,L_T^{(1)}}^{(1)} \quad \cdots \quad \mathbf{B}_{k,1}^{(P)} \ldots \mathbf{B}_{k,L_T^{(P)}}^{(P)} \right]$$

$$= \begin{bmatrix} B_{k,1,1}^{(1,1)} & \cdots & B_{k,L_T^{(1)},1}^{(1,1)} & B_{k,1,1}^{(P,1)} & \cdots & B_{k,L_T^{(P)},1}^{(P,1)} \\ \vdots & & \vdots & \vdots & & \vdots \\ B_{k,1,L_T^{(1)}}^{(1,1)} & \cdots & B_{k,L_T^{(1)},L_T^{(1)}}^{(1,1)} & B_{k,1,L_T^{(1)}}^{(P,1)} & \cdots & B_{k,L_T^{(P)},L_T^{(1)}}^{(P,1)} \\ & \vdots & & & \vdots & \\ B_{k,1,1}^{(1,P)} & \cdots & B_{k,L_T^{(1)},1}^{(1,P)} & B_{k,1,1}^{(P,P)} & \cdots & B_{k,L_T^{(P)},1}^{(P,P)} \\ \vdots & & \vdots & \vdots & & \vdots \\ B_{k,1,L_T^{(P)}}^{(1,P)} & \cdots & B_{k,L_T^{(1)},L_T^{(P)}}^{(1,P)} & B_{k,1,L_T^{(P)}}^{(P,P)} & \cdots & B_{k,L_T^{(P)},L_T^{(P)}}^{(P,P)} \end{bmatrix}. \quad (4.68)$$

As for the single-layer case (see (4.17)–(4.18)), it is assumed that

$$\hat{A}_{k,l}^{(p)} = \rho_l^{(p)} A_{k,l}^{(p)} + \Delta_{k,l}^{(p)} \quad (4.69)$$

where $E\left[\Delta_{k,l}^{(p)}\right] \approx 0$, $E\left[\Delta_{k,l}^{(p)} A_{k',l}^{(p)}\right] \approx 0$, regardless of $k$ and $k'$, and

$$E\left[\left|\Delta_{k,l}^{(p)}\right|^2\right] = \left(1 - \left(\rho_l^{(p)}\right)^2\right) E\left[\left|A_{k,l}^{(p)}\right|^2\right]. \quad (4.70)$$

The correlation coefficient $\rho_l^{(p)}$ is defined as

$$\rho_l^{(p)} = \frac{E\left[\hat{a}_{m,l}^{(p)} a_{m,l}^{(p)*}\right]}{E\left[\left|a_{m,l}^{(p)}\right|^2\right]} = \frac{E\left[\hat{A}_{k,l}^{(p)} A_{k,l}^{(p)*}\right]}{E\left[\left|A_{k,l}^{(p)}\right|^2\right]}, \quad (4.71)$$

and can be computed as described in Section 3.1.3.

In matrix notation, (4.69) takes the form

$$\hat{\mathbf{A}}_k = \mathbf{P}\mathbf{A}_k + \mathbf{\Delta}_k, \quad (4.72)$$

where the length-$N_L$ column vector

$$
\boldsymbol{\Delta}_k =
\begin{bmatrix}
\Delta_{k,1}^{(1)} \\
\vdots \\
\Delta_{k,L_T^{(1)}}^{(1)} \\
\vdots \\
\Delta_{k,1}^{(P)} \\
\vdots \\
\Delta_{k,L_T^{(P)}}^{(P)}
\end{bmatrix}
\tag{4.73}
$$

and the size-$N_L \times N_L$ matrix

$$
\mathbf{P} =
\begin{bmatrix}
\mathbf{P}^{(1)} & \cdots & 0 \\
& \ddots & \\
0 & \cdots & \mathbf{P}^{(P)}
\end{bmatrix},
\tag{4.74}
$$

with

$$
\mathbf{P}^{(p)} =
\begin{bmatrix}
\rho_1^{(p)} & \cdots & 0 \\
& \ddots & \\
0 & \cdots & \rho_{L_T^{(p)}}^{(p)}
\end{bmatrix}.
\tag{4.75}
$$

Again, by employing the Lagrange multipliers' method, it can be shown that for a SIC receiver the optimum feedforward coefficients in the MMSE sense can be written as

$$
\mathbf{F}_{k,l}^{(p)} = \left[ \mathbf{H}_k^H \left( \mathbf{I}_{N_L} - \mathbf{P}^2 \right) \mathbf{H}_k + \beta \mathbf{I}_{KL_R} \right]^{-1} \mathbf{H}_k^H \mathbf{Q}_{v(p,l)}
\tag{4.76}
$$

with

$$
\beta = \frac{E\left[ \left| N_k^{(r)} \right|^2 \right]}{E\left[ \left| A_{k,l}^{(p)} \right|^2 \right]},
\tag{4.77}
$$

(as in (2.87)), common to all $l$, $r$ and $p$, and the length-$N_L$ normalization vector

$$\mathbf{Q}_{v(p,l)} = \begin{bmatrix} 0 \\ \vdots \\ Q_{v(p,l)} \\ \vdots \\ 0 \end{bmatrix} = \left(\mathbf{I}_{N_L} - \mathbf{P}^2\right) \mathbf{\Gamma}_{v(p,l)}(k) - \frac{1}{2\sigma_A^2 M} \mathbf{\Lambda}_{v(p,l)}, \quad (4.78)$$

where $v(p, l)$ represents the position of the non-zero element associated to the $l$th layer of the $p$th MT, given by

$$v(p, l) = \begin{cases} l, & p = 1 \\ \sum_{p'=1}^{p-1} L_T^{(p')} + l, & p > 1 \end{cases}. \quad (4.79)$$

Vectors $\mathbf{\Gamma}_{v(p,l)}(k)$ and $\mathbf{\Lambda}_{v(p,l)}$ have also length $N_L$, with zeros in all positions except the $v$th, i.e.,

$$\mathbf{\Gamma}_{v(p,l)} = \begin{bmatrix} 0 \\ \vdots \\ \gamma_l^{(p)} \\ \vdots \\ 0 \end{bmatrix}, \quad (4.80)$$

where

$$\gamma_l^{(p)} = \frac{1}{M} \mathbf{F}_{k,l}^{(p)T} \mathbf{H}_{k,l}^{(p)T} = \frac{1}{M} \sum_{k=0}^{M-1} \sum_{q=0}^{K-1} \sum_{r=0}^{L_R} F_{k+qM,l}^{(p,r)} H_{k+qM,l}^{(p,r)}, \quad (4.81)$$

and

$$\mathbf{\Lambda}_{v(p,l)} = \begin{bmatrix} 0 \\ \vdots \\ \lambda_l^{(p)} \\ \vdots \\ 0 \end{bmatrix}. \quad (4.82)$$

The optimum feedback coefficients are given by

$$\mathbf{B}_{k,l}^{(p)} = \mathbf{P}\left(\mathbf{H}_k \mathbf{F}_{k,l}^{(p)} - \mathbf{\Gamma}_{v(p,l)}(k)\right). \quad (4.83)$$

Similarly, for a PIC receiver, the optimum $\mathbf{F}_k$ and $\mathbf{B}_k$ matrices are given by

$$\mathbf{F}_k = \left[\mathbf{H}_k^H \left(\mathbf{I}_{N_L} - \mathbf{P}^2\right) \mathbf{H}_k + \beta \mathbf{I}_{KL_R}\right]^{-1} \mathbf{H}_k^H \mathbf{Q}, \tag{4.84}$$

and

$$\mathbf{B}_k = \mathbf{P}\left(\mathbf{H}_k \mathbf{F}_k - \boldsymbol{\Gamma}(k)\right), \tag{4.85}$$

where

$$\mathbf{Q} = \left[\mathbf{Q}_{v(1,1)} \cdots \mathbf{Q}_{v(1,L_T^{(1)})} \quad \cdots \quad \mathbf{Q}_{v(P,1)} \cdots \mathbf{Q}_{v(P,L_T^{(P)})}\right] \tag{4.86}$$

and

$$\boldsymbol{\Lambda} = \left[\boldsymbol{\Lambda}_{v(1,1)} \cdots \boldsymbol{\Lambda}_{v(1,L_T^{(1)})} \quad \cdots \quad \boldsymbol{\Lambda}_{v(P,1)} \cdots \boldsymbol{\Lambda}_{v(P,L_T^{(P)})}\right]. \tag{4.87}$$

If we do not have data estimates for the different layers $\rho_{l'}^{(p')} = 0$ ($p' = 1, 2, \ldots, P; l' = 1, 2, \ldots, L_T^{(p')}$), and the feedback coefficients are zero. Therefore, (4.84) reduces to

$$\mathbf{F}_k = \left[\mathbf{H}_k^H \mathbf{H}_k + \beta \mathbf{I}_{KL_R}\right]^{-1} \mathbf{H}_k^H \mathbf{Q}, \tag{4.88}$$

which corresponds to a linear MUD for a MIMO system.

Extending the result of Appendix D, it can be shown that the optimum feedforward coefficients can be written in the form

$$F_{k+qM,l}^{(p,r)} = \sum_{p'=1}^{P} \sum_{l'=1}^{L_T^{(p')}} H_{k+qM,l'}^{(p',r)*} V_{k,l,l'}^{(p,p')} \tag{4.89}$$

($k = 0, 1, \ldots, M - 1; q = 0, 1, \ldots, K - 1$), with the set of coefficients $\{V_{k,l,l'}^{(p,p')}; p' = 1, \ldots, P; l' = 1, 2, \ldots, L_T^{(p')}\}$ satisfying the set of $KL_T^{(p)}$ equations

$$\sum_{p''=1}^{P} \sum_{l''=1}^{L_T^{(p'')}} V_{k,l,l''}^{(p,p'')} \left(\left(1 - \left(\rho_{l'}^{(p')}\right)^2\right) \sum_{q'=0}^{K-1} H_{k+q'M,l''}^{(p'',r)*} H_{k+q'M,l'}^{(p',r)} + \beta \delta_{l',l''} \delta_{p',p''}\right)$$
$$= \delta_{l,l'} \delta_{p,p'}, \tag{4.90}$$

$p' = 1, 2, \ldots, P; l' = 1, 2, \ldots, L_T^{(p')}$.

The computation of the feedforward coefficients from (4.89) is simpler than the direct computation, from (4.76) or (4.84), especially when $N_L < KL_R$.

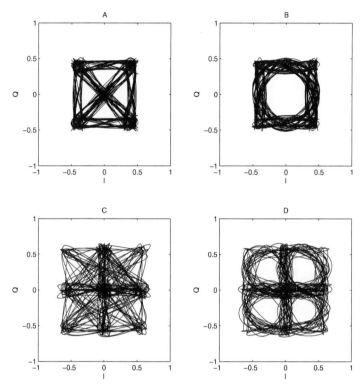

Figure 4.8 I-Q diagrams for the following situations: (A) single-code QPSK; (B) single-code OQPSK; (C) multicode QPSK; (D) multicode OQPSK.

## Implementation Issues: Multiple Transmit Antennas vs Multicode Schemes

Let us assume that the use of a single spreading code with a QPSK constellation corresponds to the data bit rate $R_b$. If we want to duplicate the bit rate while maintaining a QPSK constellation we could assign two spreading codes to a given MT, which corresponds to employing multicode CDMA schemes [68], or we could employ a space multiplexing scheme, where the MT has two antennas, each one transmitting a different data stream (naturally, this means that the BS needs two receive antennas, at least).

The major problem with multicode CDMA schemes is that the envelope fluctuations and PMEPR of the transmitted signal increase with the number of codes that is being assigned to a given MT. For instance, Fig. 4.8 shows the I-Q diagrams of the transmitted signal for a single-antenna MT with one or two spreading codes assigned to it and QPSK constellations (the PMEPR is 2.8 dB for the single-code case and 5.2 dB for the multicode case), as well as the corresponding I-Q diagrams for OQPSK schemes (Offset QPSK) (the PMEPR is 2.6 dB for the single-code case and 5.1 dB for the multicode case). A square-root raised cosine filtering with roll-off factor 0.5 is assumed. Clearly, the envelope fluctuations are much higher for the multicode scheme. The single-code case with an OQPSK scheme is of particular interest since it is compatible with a low-cost, grossly nonlinear power amplification, especially when MSK-type (Minimum Shift Keying) signals are employed.

By employing a spatial multiplexing schemes with two transmit antennas, we will need two power amplifiers; however, since the signal at the input of each amplifier is a "single-code signal", its envelope fluctuations can be very low, allowing an efficient power amplification. Moreover, the peak power required for each amplifier is lower than for the multicode case. For MTs that require very high bit rates the required number of amplifiers/antennas is also high, which is not feasible to implement. For these situations, it might be better to adopt a multicode scheme with a single amplifier and a single transmit antenna, eventually combined with some suitable signal processing for reducing the envelope fluctuations of the transmitted signals [66, 69].

It should be noted that, the different transmit and receive antennas should be almost uncorrelated. This is not a problem at the BS, since the separation between antennas can be relatively high. However, for a typical MT, which is expected to have small dimensions, this might be a problem. In this case, we could use orthogonal spreading codes for the different antennas.

Our simulations show that we can have essentially the same performance with uncorrelated antennas or highly correlated antennas, with orthogonal spreading codes. In fact, if we have a single MT, our receiver behaves as the one proposed in Chapter 3, in the second case.

It should also be noted that the separation between the data streams associated to the different antennas and the different MTs results from the combination of the spreading codes and the corresponding channel frequency responses (see (4.56)). This means that we have essentially the same performance regardless of the spreading codes, provided that we have severely time-dispersive channels and the corresponding frequency responses are highly uncorrelated. For flat fading channels and a synchronous system there is no need for a MUD receiver to separate the different data streams since the orthogonality of the corresponding spreading codes is not affected.

## Signal Processing Complexity

In the case of a MIMO system, the receiver needs $L_R$ size-$N$ DFT operations (one for each receive antenna), a pair of size-$M$ DFT/IDFT operations (except for the first iteration where only a size-$M$ IDFT is required) and a decision/decoding operation for the detection of each layer, at each iteration. The computation of the feedforward coefficients requires inverting the $KL_R \times KL_R$ matrix in (4.84), or solving a system of $N_L$ equations, for each frequency, if (4.89)–(4.90) are used. Whenever a given layer has a very high reliability ($\rho_l^{(p)} \approx 1$), it can be ignored when detecting the other layers in the next iterations; therefore the computation of the feedforward coefficients requires solving a system with a smaller dimension, provided that (4.89)–(4.90) are used. Again, the most complex part of the receiver's implementation is when the SISO channel decoding needs to be implemented for the detection of each layer, at each iteration, in the case of the Turbo MUD-SD receiver.

### 4.2.4 Performance Results

In this section, we present a set of performance results concerning the proposed SIC and PIC receivers with joint turbo equalization and multiuser detection. We consider the uplink transmission within a CP-assisted DS-CDMA system with spreading factor $K = 4$ and $M = 256$ data symbols for each user, corresponding to blocks with length $N = KM = 1024$, plus an appropriate CP.

Let us first consider a single-layer system and a single-user scenario ($P = 1$). Figure 4.9 shows the impact of the spreading factor ($K = 1$, 4 and 16) on the uncoded BER performance. For the sake of comparisons, we also include the corresponding MFB performance. From this figure, we can observe that the iterative receiver structure allows a significant improvement on the BER performance, especially for low spreading factors. We can also see that the performance approaches asymptotically the MFB. For higher values of $K$ the performance is better due to the implicit diversity inherent to the spreading operation (although the MFB remains constant since we fixed $N$ and the bandwidth). For $K = 1$ (i.e., without spreading), the first iteration corresponds to the linear FDE and the remaining iterations corresponds to the IB-DFE receiver for CP-assisted SC modulations.

Let us consider now the case when $P = K = 4$ (i.e., a fully loaded system) where the signals associated to all users have the same average power at the receiver (i.e., the BS), which corresponds to a scenario where an "ideal average power control" is implemented. Figure 4.10 shows the uncoded BER

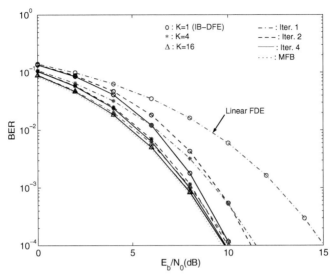

Figure 4.9  Uncoded BER performance for an SU scenario when $K = 1, 4$ or 16 and a receiver with 1, 2 or 4 iterations.

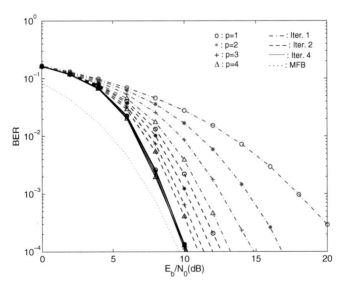

Figure 4.10 Uncoded BER performance for each user and a SIC receiver with 1, 2 or 4 iterations, when $P = K = 4$ (for a given iteration, the users that are detected later have better BER).

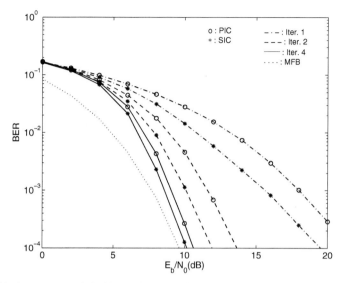

Figure 4.11  Average uncoded BER performance for each iteration for PIC and SIC receivers, when $P = K = 4$.

performance for the IMUD-HD as well as the corresponding MFB perform-ance. We can observe that the iterative receiver structure allows a significant improvement on the BER performance. Since a SIC scheme was considered, for a given iteration the users that are detected first face stronger interference levels and have worse BER. After four iterations the performance is already similar for all users, and asymptotically close to the MFB; however, for low SNR we are still far from the MFB. Figure 4.11 presents the average uncoded BER for each iteration (averaged over all the users) for PIC and SIC receivers. From this figure, it is clear that the PIC receiver has worse performance, especially for the first iteration. This is a consequence of the fact that the PIC receiver uses the block estimates from the previous iteration for all users and the SIC receiver uses the most updated version of it. After four iterations the performance of PIC and SIC receivers are similar and, once again, very close to the MFB.

Let us consider now a scenario where the signals associated to different users have different average power at the receiver. We will consider two classes of users, $C_L$ and $C_H$, with different average powers at the receiver. The performance results presented in Fig. 4.12 concern the case when the average power of $C_H$ users is 6 dB above the average power of $C_L$ users. The system is fully loaded, with two $C_L$ users and two $C_H$ users, and a

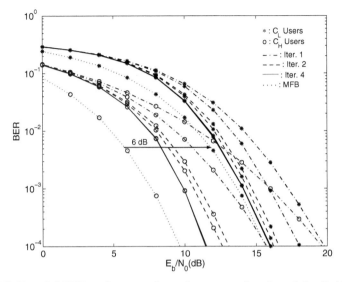

Figure 4.12 Uncoded BER performance for each user as a function of the $E_b/N_0$ of $C_H$ users, when $P = K = 4$.

SIC scheme. Clearly, the $C_L$ users face strong interference conditions. Once again, the proposed iterative receiver allows significant performance gains. In both cases, the performance of $C_L$ users asymptotically approaches the MFB when we increase the number of iterations; however, for $C_H$ users, the BER at $10^{-4}$ is still between 1 or 2 dB from the MFB. This can be explained from the fact that the BER is much lower for $C_H$ users, allowing an almost perfect interference cancelation of their effects on $C_L$ users; therefore, the corresponding performance can be very close to the MFB. The higher BERs for the $C_L$ users preclude an appropriate interference cancelation when we detect $C_H$ users (similar results were obtained for a PIC receiver).

In Fig. 4.13 we compare the average uncoded performance for each iteration (again, averaged over all the users) for a SIC receiver when either hard decisions or soft decisions are used in the feedback loop, i.e., for the IMUD-HD and the IMUD-SD, respectively. Clearly, the use of soft decisions allows just a slight performance improvement (a few tens of a dB, at most). Similar results were observed for the PIC receiver.

Let us consider now the impact of channel coding. Figure 4.14 is concerned, once again, with a fully loaded system ($P = K = 4$), in the following cases: feedback with soft decisions from the IMUD (IMUD-SD); coded feedback with a conventional Viterbi decoder, assuming $\rho_p = \rho_{m,p}^I = \rho_{m,p}^Q = 1$ (Turbo MUD-HD); coded feedback with a SISO decoder, implemented using

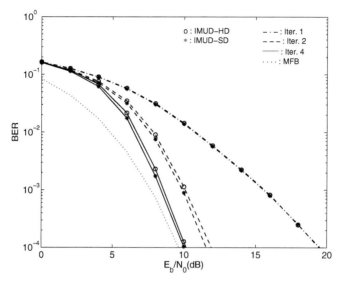

Figure 4.13 Average uncoded BER performance for each iteration and a SIC receiver employing either hard decisions and soft decisions in the feedback loop, when $P = K = 4$.

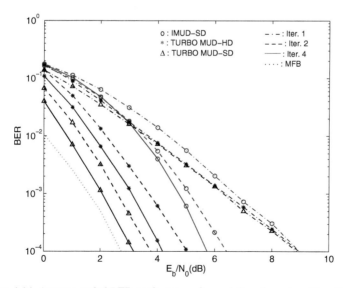

Figure 4.14 Average coded BER performance for each iteration when $P = K = 4$.

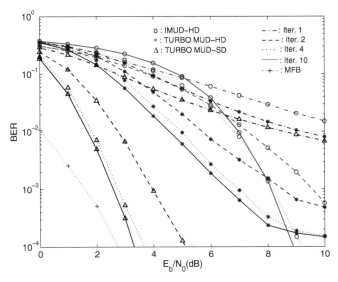

Figure 4.15 Average coded BER performance for each iteration when $P = 6$ and $K = 4$.

the Max-Log-MAP approach (Turbo MUD-SD). From this figure, it is clear that the turbo receivers (Turbo MUD-HD and Turbo MUD-SD) outperform the receiver that uses soft decisions from the IMUD in the feedback loop (IMUD-SD), even the suboptimum Turbo MUD-HD based on the Viterbi decoder. Moreover, the performance of Turbo MUD-SD approaches the corresponding MFB, contrarily to the other schemes. Let us consider now an overloaded scenario, i.e., when $P > K$. In Fig. 4.15 we show the average performance for the receivers of Fig. 4.14, when $P = 6$. In this case, the receiver is able to separate the different users, even when using an uncoded feedback, although the required number of iterations is higher and with some performance degradation.

If we increase the number of users the only receiver able to separate them is the turbo receiver that uses the SISO block in the feedback loop. For instance, Fig. 4.16 shows the average performance when $P = 8$, i.e., the number of users is twice the spreading factor.

It should be pointed out that similar results could be obtained for other severely time-dispersive channels. For instance, Figs. 4.17 and 4.18 show the coded BER performance for the same conditions of Fig. 4.14, but with two different severely time-dispersive channels: Fig. 4.17 concerns a channel with uncorrelated Rayleigh fading on the different frequencies and Fig. 4.18 con-

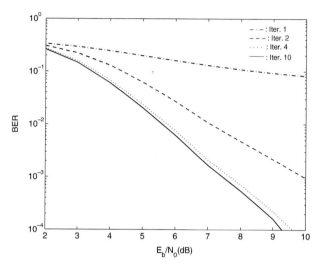

Figure 4.16 Average coded BER performance for each iteration with the Turbo MUD-SD receiver when $P = 8$ and $K = 4$.

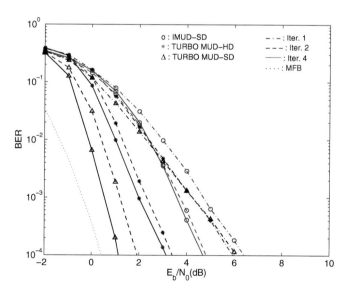

Figure 4.17 Average coded BER performance as in Fig. 4.14 for a channel with uncorrelated Rayleigh fading on the different frequencies.

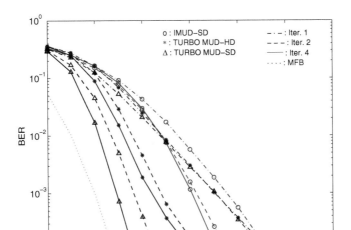

Figure 4.18 Average coded BER performance as in Fig. 4.14 for an exponential PDP channel with uncorrelated Rayleigh fading on the different paths.

cerns an exponential PDP channel with uncorrelated Rayleigh fading on the different paths (see Fig. A.2). Clearly, the same conclusions can be derived as in Fig. 4.14.

Let us consider now a MIMO system employing spatial multiplexing and a random spreading with spreading factor $K = 4$ and $M = 256$ data symbols for each layer, corresponding to blocks with length $N = KM = 1024$. It is assumed that the BS has $L_R = 2$ uncorrelated receive antennas and a fully loaded scenario with $P = 4$ MTs, each one with $L_T^{(p)} = 2$ uncorrelated transmit antennas (i.e., a spatial multiplexing scheme with two layers per MT). The signals associated to each antenna have the same average power at the receiver (i.e., the BS), which corresponds to a scenario where an "ideal average power control" is implemented. The performance results presented in the following concern to the IMUD-SD and a SIC receiver, although similar results are obtained with a PIC receiver.

Figures 4.19 to 4.22 show the impact of the number of iterations on the BER for each layer of $MT_1$ to $MT_4$, respectively. For the sake of comparisons, we also include the corresponding MFB performance. Figure 4.23 shows the average BER for each MT (i.e., the average over the two transmitted layers). From these figures, we can observe that our iterative receiver structure allows

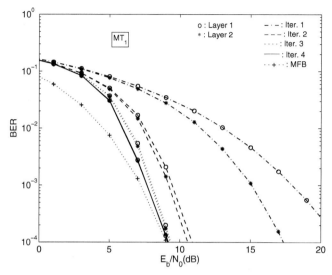

Figure 4.19 BER performance for layer $l$ ($l = 1, 2$) of $MT_1$ and iterations 1 to 4 when the average received power is identical for all MTs (layers detected latter have better performance, i.e., the worse performance is for $(i, l)=(1,1)$, followed by $(1,2)$, $(2,1)$, $(2,2)$, $(3,1)$, $(3,2)$, $(4,1)$ and $(4,2)$).

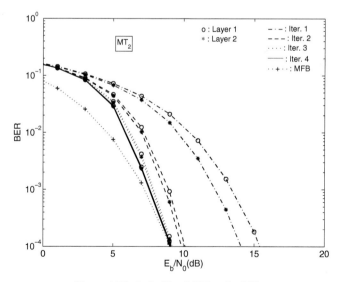

Figure 4.20 As in Fig. 4.19 but for $MT_2$.

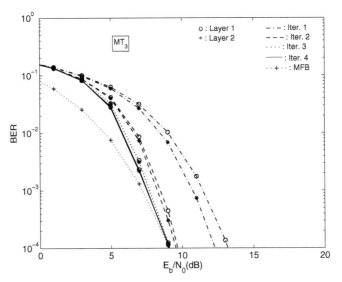

Figure 4.21  As in Fig. 4.19 but for $MT_3$.

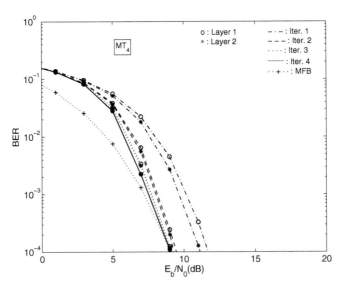

Figure 4.22  As in Fig. 4.19 but for $MT_4$.

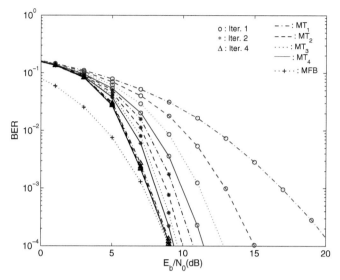

Figure 4.23 Average uncoded BER performance for each MT and 1, 2 or 4 iterations, when the average received power is identical for all MTs.

a significant improvement on the BER performance. As with single-layer systems (see Fig. 4.10), for a given iteration, the layers that are detected first face stronger interference levels and have worse BER. This is especially important at the first iteration. After four iterations the performance is already similar for all layers, and very close to the MFB.

Let us assume now that we have different average receive powers for the different MTs. We will assume that the difference between the average receive power of $MT_1$ and the average receive power of $MT_2$, $MT_3$ and $MT_4$ are 3 dB, 6 dB and 9 dB, respectively. Clearly, the MTs with higher $p$ face stronger interference levels. The corresponding layers' performance is depicted in Figs. 4.24 to 4.27. Once again, the proposed iterative receiver allows significant performance gains. The performance of MTs with lower power asymptotically approaches the MFB when we increase the number of iterations; however, for MTs with higher power, the BER at $10^{-4}$ is still between 1 or 2 dB from the MFB. This can be explained from the fact that the BER is much lower for high-power users, allowing an almost perfect interference cancelation of their effects on low-power users; therefore, the corresponding performance can be very close to the MFB. The higher BERs for the low-power users preclude an appropriate interference cancelation when we detect high-power users (see also Fig. 4.28).

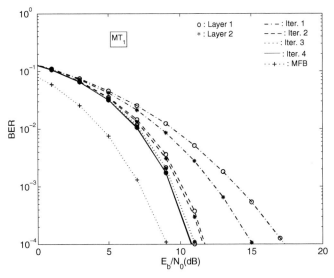

Figure 4.24 BER performance for layer $l$ ($l = 1, 2$) of $MT_1$ and iterations 1 to 4 when the different MTs have different average receive powers (average received powers of $MT_2$, $MT_3$ and $MT_4$ are 3 dB, 6 dB and 9 dB below the average power of $MT_1$, respectively.

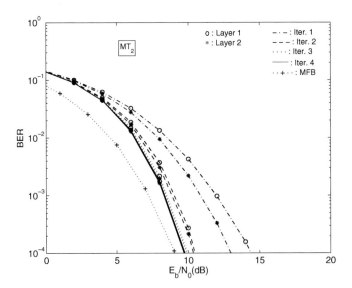

Figure 4.25  As in Fig. 4.24 but for $MT_2$.

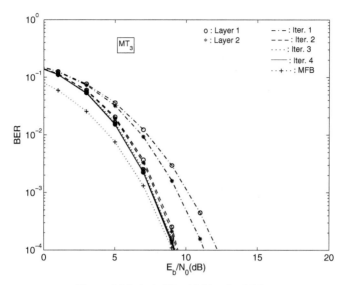

Figure 4.26  As in Fig. 4.24 but for $MT_3$.

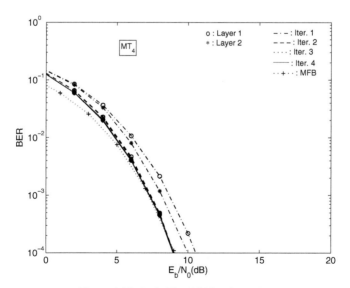

Figure 4.27  As in Fig. 4.24 but for $MT_4$.

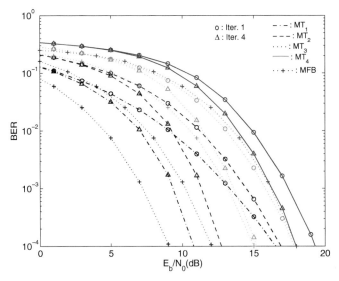

Figure 4.28 Average uncoded BER performance for each MT and 1 or 4 iterations, expressed as a function of the $E_b/N_0$ of $MT_1$.

## 4.3 Receivers for the Uplink Transmissions of a MC-CDMA System

### 4.3.1 System Characterization

Let us consider the uplink transmission of MC-CDMA signals employing frequency-domain spreading. The generic system's architecture is depicted in Fig. 4.29 and corresponds to a MIMO system with $P$ users (MTs), transmitting independent data blocks with the same dimensions, and $L$ receive antennas at the BS. For the sake of simplicity, it is assumed that all users have the same spreading factor $K$ and the same data rate.

In this section we will assume that each MT has a single transmit antenna (the extension for multiple transmit antennas follows very closely the spacial multiplexing techniques presented in Section 4.2.3 for DS-CDMA systems). The coded bits are interleaved and mapped, leading to the size-$M$ data symbols block to be transmitted by the $p$th MT $\{A_{m,p}; k = 0, 1, \ldots, M - 1\}$. The frequency-domain block of chips to be transmitted by the $p$th MT is $\{S_{k,p}; k = 0, 1, \ldots, N - 1\}$, with $S_{k,p}$ given by (2.63), i.e.,

$$S_{k,p} = C_{k,p} A_{k \bmod M,p}. \tag{4.91}$$

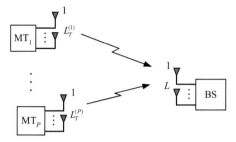

Figure 4.29 MC-CDMA system characterization.

## 4.3.2 Receiver Structure

As usual, it is assumed that the length of the CP is higher than the length of the overall channel impulse response. We will assume that the BS has $L$ receive antennas and the received time-domain block associated to the $l$th diversity branch, after discarding the samples associated to the CP, is $\{y_n^{(l)}; n = 0, 1, \ldots, N-1\}$. The corresponding frequency-domain block (i.e., the length-$N$ DFT of the block $\{y_n^{(l)}; n = 0, 1, \ldots, N-1\}$) is $\{Y_k^{(l)}; k = 0, 1, \ldots, N-1\}$, with

$$
\begin{aligned}
Y_k^{(l)} &= \sum_{p=1}^{P} S_{k,p} \, H_{k,p}^{Ch(l)} + N_k^{(l)} \\
&= \sum_{p=1}^{P} A_{k \bmod M, p} \, C_{k,p} \, \xi_p \, H_{k,p}^{Ch(l)} + N_k^{(l)} \\
&= \sum_{p=1}^{P} A_{k \bmod M, p} \, H_{k,p}^{(l)} + N_k^{(l)}
\end{aligned}
\tag{4.92}
$$

with $H_{k,p}^{Ch(l)}$ denoting the channel frequency response between the $p$th MT and the $l$th diversity branch, at the $k$th subcarrier, $N_k^{(l)}$ the corresponding channel noise and

$$
H_{k,p}^{(l)} = \xi_p H_{k,p}^{Ch(l)} C_{k,p}.
\tag{4.93}
$$

As it is known, the detection of the $m$th symbol of the $p$th MT involves the set of subcarriers $\Psi_m = \{m, m + M, \ldots, m + (K-1)M\}$ used to transmit that symbol.

By defining the length-$KL$ column vector

$$\mathbf{Y}(k) = \begin{bmatrix} \mathbf{Y}^{(1)}(k) \\ \vdots \\ \mathbf{Y}^{(L)}(k) \end{bmatrix}, \tag{4.94}$$

with

$$\mathbf{Y}^{(l)}(k) = \begin{bmatrix} Y_k^{(l)} \\ \vdots \\ Y_{k+(K-1)M}^{(l)} \end{bmatrix} \tag{4.95}$$

denoting the length-$K$ column vector with the received samples associated to the set of frequencies $\Psi_m$, for the $l$th antenna, (4.92) can be written as

$$\mathbf{Y}(k) = \mathbf{H}^T(k)\mathbf{A}(k) + \mathbf{N}(k), \tag{4.96}$$

where

$$\mathbf{N}(k) = \begin{bmatrix} \mathbf{N}^{(1)}(k) \\ \vdots \\ \mathbf{N}^{(L)}(k) \end{bmatrix}, \tag{4.97}$$

with

$$\mathbf{N}^{(l)}(k) = \begin{bmatrix} N_k^{(l)} \\ \vdots \\ N_{k+(K-1)M}^{(l)} \end{bmatrix} \tag{4.98}$$

denoting the vector with the noise samples associated to the set of frequencies $\Psi_m$, for the $l$th antenna. In (4.96), $\mathbf{H}(k)$ is the $P \times KL$ overall channel frequency response matrix associated to $\mathbf{A}(k)$, i.e.,

$$\mathbf{H}(k) = \begin{bmatrix} H_{k,1}^{(1)} & \cdots & H_{k+(K-1)M,1}^{(1)} & H_{k,1}^{(L)} & \cdots & H_{k+(K-1)M,1}^{(L)} \\ \vdots & & \vdots & \cdots & \vdots & & \vdots \\ H_{k,P}^{(1)} & \cdots & H_{k+(K-1)M,P}^{(1)} & H_{k,P}^{(L)} & \cdots & H_{k+(K-1)M,P}^{(L)} \end{bmatrix}$$

$$= \begin{bmatrix} \mathbf{H}^{(1)}(k) & \cdots & \mathbf{H}^{(L)}(k) \end{bmatrix} = \begin{bmatrix} \mathbf{H}_1(k) \\ \vdots \\ \mathbf{H}_P(k) \end{bmatrix}, \tag{4.99}$$

where

$$\mathbf{H}^{(l)}(k) = \begin{bmatrix} H_{k,1}^{(l)} & \cdots & H_{k+(K-1)M,1}^{(l)} \\ \vdots & & \vdots \\ H_{k,P}^{(l)} & \cdots & H_{k+(K-1)M,P}^{(l)} \end{bmatrix} \tag{4.100}$$

(a)

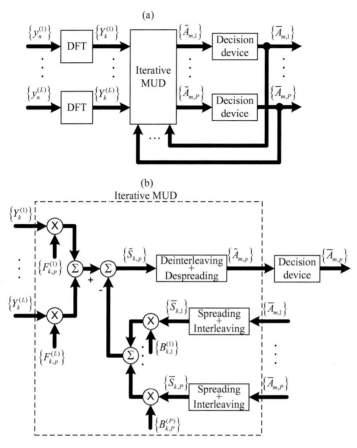

(b)
Iterative MUD

Figure 4.30 (a) Iterative receiver for a linear transmitter and (b) detail of the detection of the $p$th MT.

is a $P \times K$ matrix with lines associated to the different MTs and columns associated to the set of frequencies $\Psi_m$, for the $l$th antenna, and

$$\mathbf{H}_p(k) = \left[ \begin{array}{ccccccc} H_{k,p}^{(1)} & \cdots & H_{k+(K-1)M,p}^{(1)} & \cdots & H_{k,p}^{(L)} & \cdots & H_{k+(K-1)M,p}^{(L)} \end{array} \right]$$
(4.101)

is the length-$KL$ line vector with the channel frequency response coefficients associated to the $p$th MT.

This receiver can be regarded as an iterative MUD receiver, as depicted in Fig. 4.30(a). The receiver with "soft decisions" in the feedback will only be considered since it leads to better performance than the receiver with "hard decisions", as we saw in the previous section. For a given iteration, the detection of $\mathbf{A}(k)$ employs the structure depicted in Fig. 4.30(b), where we have $L$ feedforward filters (one for each receive antennas) and $P$ feedback loops. As before, the feedforward filters are designed to minimize the MAI that cannot be canceled by the feedback loops. For the first iteration no information exists on the MT's transmitted symbols and the receiver reduces to a linear MUD.

For each iteration, the equalized samples vector $\tilde{\mathbf{A}}(k)$, associated to $\mathbf{A}(k)$, is given by

$$\tilde{\mathbf{A}}(k) = \mathbf{F}^T(k)\mathbf{Y}(k) - \mathbf{B}^T(k)\overline{\mathbf{A}}(k) \tag{4.102}$$

where $\mathbf{F}(k)$ is a $KL \times P$ matrix with the feedforward coefficients given by

$$\mathbf{F}(k) = \begin{bmatrix} F_{k,1}^{(1)} & \cdots & F_{k,P}^{(1)} \\ \vdots & & \vdots \\ F_{k+(K-1)M,1}^{(1)} & \cdots & F_{k+(K-1)M,P}^{(1)} \\ & \vdots & \\ F_{k,1}^{(L)} & \cdots & F_{k,P}^{(L)} \\ & \vdots & \\ F_{k+(K-1)M,1}^{(L)} & \cdots & F_{k+(K-1)M,P}^{(L)} \end{bmatrix} = \begin{bmatrix} \mathbf{F}^{(1)}(k) \\ \vdots \\ \mathbf{F}^{(L)}(k) \end{bmatrix}$$

$$= \begin{bmatrix} \mathbf{F}_1(k) & \cdots & \mathbf{F}_P(k) \end{bmatrix}, \tag{4.103}$$

where

$$\mathbf{F}^{(l)}(k) = \begin{bmatrix} F_{k,1}^{(l)} & \cdots & F_{k,P}^{(l)} \\ \vdots & & \vdots \\ F_{k+(K-1)M,1}^{(l)} & \cdots & F_{k+(K-1)M,P}^{(l)} \end{bmatrix} \tag{4.104}$$

and

$$\mathbf{F}_p(k) = \begin{bmatrix} F_{k,p}^{(1)} \\ \vdots \\ F_{k+(K-1)M,p}^{(1)} \\ \vdots \\ F_{k,p}^{(L)} \\ \vdots \\ F_{k+(K-1)M,p}^{(L)} \end{bmatrix}, \tag{4.105}$$

and $\mathbf{B}(k)$ is the $P \times P$ matrix with the feedback coefficients given by

$$
\mathbf{B}(k) = \begin{bmatrix} B_{k,1}^{(1)} & \cdots & B_{k,P}^{(1)} \\ \vdots & & \vdots \\ B_{k,1}^{(P)} & \cdots & B_{k,P}^{(P)} \end{bmatrix} = \begin{bmatrix} \mathbf{B}_1(k) & \cdots & \mathbf{B}_P(k) \end{bmatrix} = \begin{bmatrix} \mathbf{B}^{(1)}(k) \\ \vdots \\ \mathbf{B}^{(P)}(k) \end{bmatrix},
$$

$$(4.106)$$

with

$$
\mathbf{B}_p(k) = \begin{bmatrix} B_{k,p}^{(1)} \\ \vdots \\ B_{k,p}^{(P)} \end{bmatrix} \qquad (4.107)
$$

and

$$
\mathbf{B}^{(p)}(k) = \begin{bmatrix} B_{k,1}^{(p)} & \cdots & B_{k,P}^{(p)} \end{bmatrix}. \qquad (4.108)
$$

$\overline{\mathbf{A}}(k)$ is the vector with the "soft decisions" of $\mathbf{A}(k)$, given by (3.46), from the multiuser detector, obtained at the previous iteration. As in the uplink transmission of DS-CDMA scheme a turbo receiver can also be defined by employing the "soft decisions" from the SISO channel decoder outputs (Turbo MUD-SD) instead of the "soft decisions" from the multiuser detector (IMUD-SD). The input of the SISO block are the LLRs of the "coded bits" at the output of the multiuser detector given by (3.43) and (3.44).

### Derivation of the Receiver Parameters

Once again, as with (4.17), it is assumed that

$$
\hat{A}_{m,p} = \rho_p A_{m,p} + \Delta_{m,p} \qquad (4.109)
$$

and

$$
\overline{A}_{m,p} = \rho_p \hat{A}_{m,p} = \rho_p^2 A_{m,p} + \rho_p \Delta_{m,p}. \qquad (4.110)
$$

In matrix notation, (4.110) takes the form

$$
\overline{\mathbf{A}}(k) = \mathbf{P}^2 \mathbf{A}(k) + \mathbf{P}\mathbf{\Delta}(k), \qquad (4.111)
$$

with

$$
\mathbf{\Delta}(k) = \begin{bmatrix} \Delta_{m,1} \\ \vdots \\ \Delta_{m,P} \end{bmatrix} \qquad (4.112)
$$

where $\Delta_{m,p}$ denotes the error associated to the $m$th symbol of the $p$th MT.

By combining (4.96), (4.102) and (4.111), it leads to

$$\tilde{\mathbf{A}}(k) = \underbrace{\mathbf{\Gamma}(k)\mathbf{A}(k)}_{\text{Useful signal}} + \underbrace{\left(\mathbf{F}^T(k)\,\mathbf{H}^T(k) - \mathbf{B}^T(k)\,\mathbf{P}^2 - \mathbf{\Gamma}(k)\right)\mathbf{A}(k)}_{\text{Residual Interference (ISI+MAI)}}$$

$$- \underbrace{\mathbf{B}^T(k)\mathbf{P}\mathbf{\Delta}(k)}_{\substack{\text{"Noise" due to} \\ \text{feedback errors}}} + \underbrace{\mathbf{F}^T(k)\,\mathbf{N}(k)}_{\text{Channel noise}}. \tag{4.113}$$

Choosing $\mathbf{F}(k)$ and $\mathbf{B}(k)$ so as to maximize

$$\text{SINR}_p = \frac{E\left[|A_{m,p}|^2\right]}{E\left[|\Theta_{m,p}|^2\right]} \tag{4.114}$$

for $p = 1, ..., P$, at a particular iteration, where

$$\Theta_{m,p} = \tilde{A}_{m,p} - A_{m,p} \tag{4.115}$$

denotes the overall error for the $m$th symbol of the $p$th MT, is equivalent to minimize the MSE

$$\begin{aligned}
E\left[|\mathbf{\Theta}(k)|^2\right] &= E\left[\left|\tilde{\mathbf{A}}(k) - \mathbf{A}(k)\right|^2\right] \\
&= E\left[\left|\left(\mathbf{F}^T(k)\mathbf{H}^T(k) - \mathbf{B}^T(k)\mathbf{P}^2 - \mathbf{\Gamma}(k)\right)\mathbf{A}(k)\right|^2\right] \\
&\quad + E\left[\left|\mathbf{B}^T(k)\mathbf{P}\mathbf{\Delta}(k)\right|^2\right] + E\left[\left|\mathbf{F}^T(k)\mathbf{N}(k)\right|^2\right]
\end{aligned} \tag{4.116}$$

conditioned to $\mathbf{\Gamma}(k) = \mathbf{I}_P$.

Once again, by defining the matrix of Lagrange function

$$\mathbf{J} = E\left[|\mathbf{\Theta}(k)|^2\right] + (\mathbf{\Gamma}(k) - \mathbf{I}_P)\,\mathbf{\Lambda}, \tag{4.117}$$

and assuming that the optimization is carried out under $\mathbf{\Gamma}(k) = \mathbf{I}_P$ the optimum $\mathbf{F}(k)$ and $\mathbf{B}(k)$ matrices are obtained by solving the following set of equations:

- $\nabla_{\mathbf{F}(k)}\mathbf{J} = 0 \Leftrightarrow \mathbf{H}^H(k)\mathbf{R}_A\mathbf{H}(k)\mathbf{F}(k) - \mathbf{H}^H(k)\mathbf{R}_A\mathbf{P}^2\mathbf{B}(k) - \mathbf{H}^H(k)\mathbf{R}_A$

$$+ \mathbf{R}_N\mathbf{F}(k) + \frac{1}{M}\mathbf{H}^H(k)\mathbf{\Lambda} = 0$$

$$\Leftrightarrow \mathbf{H}^H(k)\mathbf{H}(k)\mathbf{F}(k) - \mathbf{H}^H(k)\mathbf{P}^2\mathbf{B}(k) - \mathbf{H}^H(k)$$

$$+ \beta\mathbf{F}(k) + \frac{1}{2M\sigma_A^2}\mathbf{H}^H(k)\mathbf{\Lambda} = 0, \tag{4.118}$$

with $\mathbf{R}_A = E\left[\mathbf{A}^*(k)\mathbf{A}^T(k)\right] = 2\sigma_A^2\mathbf{I}_P$, $\mathbf{R}_N = E\left[\mathbf{N}^*(k)\mathbf{N}^T(k)\right] = 2\sigma_N^2\mathbf{I}_{KL}$;

- $\nabla_{\mathbf{B}(k)}J = 0 \Leftrightarrow \left(\mathbf{P}^2\mathbf{R}_A + \mathbf{R}_\Delta\right)\mathbf{B}(k) = \mathbf{R}_A\mathbf{H}(k)\mathbf{F}(k) - \mathbf{R}_A$; $\hspace{1em}$ (4.119)

- $\nabla_{\mathbf{A}}J = 0 \Leftrightarrow \mathbf{\Gamma}(k) = \mathbf{I}_P$. $\hspace{1em}$ (4.120)

From (4.119), the optimum feedback matrix is

$$\mathbf{B}(k) = \mathbf{H}(k)\mathbf{F}(k) - \mathbf{I}_P \hspace{2em} (4.121)$$

and replacing (4.121) in (4.118), the optimum feedforward matrix is

$$\mathbf{F}(k) = \left[\mathbf{H}^H(k)\left(\mathbf{I}_P - \mathbf{P}^2\right)\mathbf{H}(k) + \beta\mathbf{I}_{KL}\right]^{-1}\mathbf{H}^H(k)\mathbf{Q}. \hspace{1em} (4.122)$$

For the first iteration, $\mathbf{P} = \mathbf{B}(k) = 0$ and (4.122) reduces to a linear MUD receiver presented in Chapter 2.

The optimum feedforward matrix (4.122) can also be written as (4.42), which allow a simpler computation, especially when $P < KL$.

### Signal Processing Complexity

To implement the IMUD receiver for linear transmitters we need $L$ size-$N$ DFT operations, one for each antenna, and a pair of despreading/spreading operations for the detection of each MT, at each iteration (except for the first iteration where only one despreading operation for each MT is required). As for the computation of the feedforward coefficients, we need to invert the $P \times P$ matrix of (4.42) for each MT, at each iteration, unless in the case of slow-varying channels where this operation is not required for all blocks. In the case of the Turbo MUD receiver the SISO channel decoding needs to be implemented in the detection process of each MT. As in the case of the Turbo MUD receiver for the DS-CDMA system, this can be the most complex part of the Turbo MUD receiver. Nevertheless, it should be pointed out that the implementation charge is concentrated in the BS, where increased power consumption and cost are not so critical.

Once again, whenever $\rho_p \approx 1$ for the $p$th MT at a given iteration it can be excluded from the detection process in the next iterations, decreasing the receiver's implementation charge.

### 4.3.3 Performance Results

This section presents a set of performance results concerning the proposed MUD receivers for the uplink transmission of MC-CDMA systems. It is

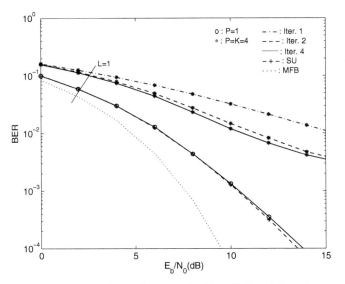

Figure 4.31 Average uncoded BER performance for $N = KM = 1024$, $K = 4$ and $P = 1$ and 4 users for iterations 1, 2 and 4.

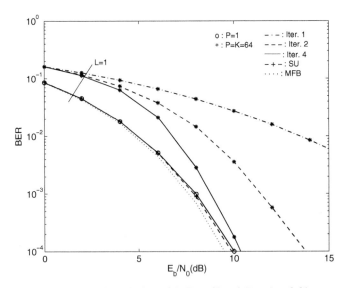

Figure 4.32 As in Fig. 4.31 but with $K = 64$ and $P = 1$ and 64 users.

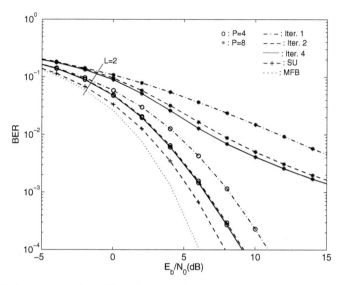

Figure 4.33 Average uncoded BER performance for $K = 4$ and $P = 4$ and 8 users with a two branch space diversity receiver ($L = 2$).

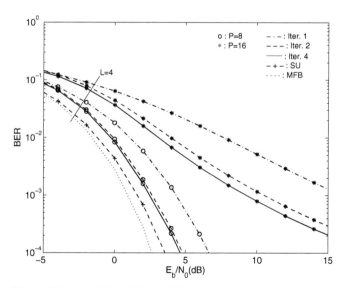

Figure 4.34 As in Fig. 4.33 but with $L = 4$ and $P = 8$ and 16 users.

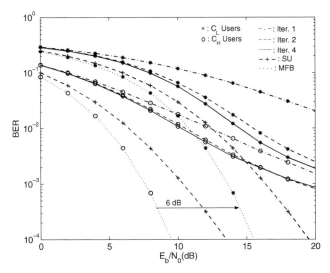

Figure 4.35 Average uncoded BER performance for $K/2 = 2\,C_L$ users and $K/2 = 2\,C_H$ users.

assumed that each MT transmits blocks of length $N = KM = 1024$ sub-carriers, plus an appropriate CP, employing a linear transmitter and that the receiver has $L$ receive antennas for diversity purposes. The performance results concern only to receivers with a PIC structure using soft decisions in the feedback loop (i.e., IMUD-SD and Turbo MUD-SD receivers) and the same severely time-dispersive channel of Section 1.3 (as with the DS-CDMA scheme, similar conclusions can be derived for other severely time-dispersive channels).

Let us start with an uncoded case where all users have the same average receive power (i.e., a perfect power control) and a single receive antenna at the receiver. Figures 4.31 and 4.32 show the uncoded BER performance for iterations 1, 2 and 4 for two fully loaded scenarios, e.g., $P = K = 4$ and $P = K = 64$ users, respectively. The SU performance ($P = 1$), which is optimum, is also shown in both cases. From these figures we can observe, once again, that the iterative procedure allows significant performance improvements relatively to the linear receiver, especially for high values of $K$. In fact, as was already mentioned in Chapter 3, the higher is the diversity effect on the transmitted symbols in the frequency-domain inherent to the spreading procedure, the closer is the performance to the corresponding SU and MFB performance. As shown in Figs. 4.33 and 4.34, if we increase the

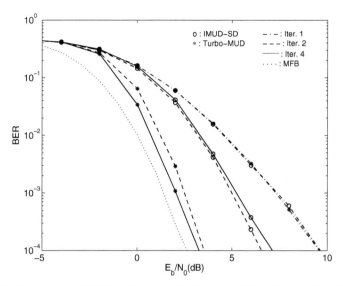

Figure 4.36 Average coded BER performance for iterations 1, 2 and 4 for either IMUD-SD and Turbo MUD-SD receivers with $L = 1$ and $P = K = 4$.

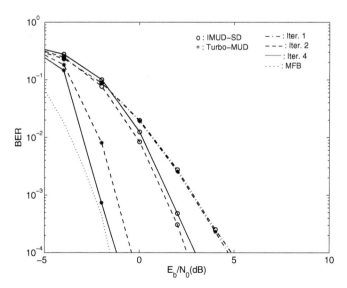

Figure 4.37 As in Fig. 4.36 but with $L = 2$ and $P = LK = 8$.

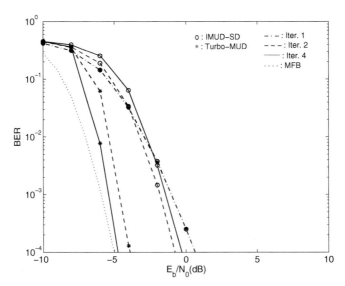

Figure 4.38  As in Fig. 4.36 but with $L = 4$ and $P = LK = 16$.

number of receive antennas we can compensate the worse performance for small spreading factors. However, when $P = KL$ the performance remains far from the SU bound even after four iterations.

When different powers are assigned to different users, as shown in Fig. 4.35, where $P = K = 4$, with two $C_L$ users and two $C_H$ users with more 6 dB of average power than $C_L$ users, we can observe that the performance of $C_H$ is further apart from the corresponding SU performance than the one of $C_L$ users for the reasons that was already explained, i.e., the higher BERs of low-power users preclude an efficient cancelation of their interference when high-power users are detected.

Let us consider a coded transmission, again with a small spreading factor of $K = 4$. Figures 4.36 to 4.38 show the average coded BER performance for iterations 1, 2 and 4, for the IMUD and Turbo MUD receivers, with $L = 1, 2$ and 4 receive antennas, respectively. Once again, the channel coding allows significant performance improvements relatively to the uncoded case. As expected, the Turbo MUD receiver out preforms the IMUD receiver. Moreover, the coded BER performance or the IMUD receiver tends to get worse from iteration to iteration due to error propagation effects inherent to highly interference levels. When we increase the spreading factor, as shown in Fig. 4.39, where $P = K = 64$ and $L = 1$, not even the performance improve for

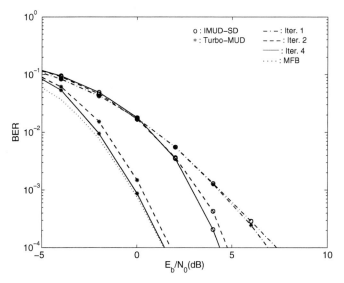

Figure 4.39  As in Fig. 4.36 but with $P = K = 64$.

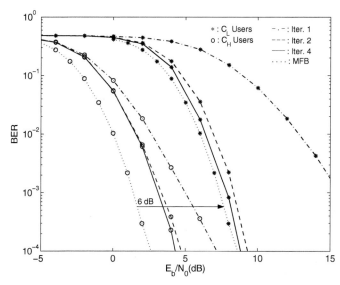

Figure 4.40  Average coded BER performance for the Turbo MUD receiver, with $K/2 = 2$ $C_L$ users and $K/2 = 2$ $C_H$ users.

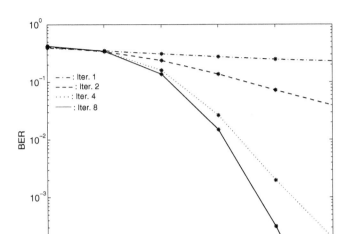

Figure 4.41  Average coded BER performance for the Turbo MUD receiver for an overloaded scenario, with $P = 2K = 8$ users and $L = 1$.

both receivers, especially at low and moderate SNR, as the error propagation effects of the IMUD receiver tends to vanish.

Figure 4.40 shows the average coded BER performance for the Turbo MUD receiver for the scenario where users have different powers, as in Fig. 4.35. Clearly, after four iterations the performance is very close to the MFB, especially for $C_L$ users, which benefit from an appropriate cancelation of the effects of $C_H$ users in the detection process.

Finally, let us consider an overloaded scenario, with $P = 8$ users, i.e., the number of users is twice the spreading factor, and the Turbo MUD receiver. As we can see from Fig. 4.41, the receiver is still able to separate the different users, although with some performance degradation and a higher number of iterations.

# 5

# DS-CDMA and MC-CDMA Systems with Nonlinear Transmitters

One of the major problems with MC-CDMA signals is the high envelope fluctuations and high PMEPR values which, as with other multicarrier modulations, can lead to amplification difficulties since, typically, a single power amplifier has to deal with a signal having a large PMEPR. The same problem occurs in the downlink transmission of DS-CDMA signals as well as for any multicode DS-CDMA scheme since the transmitted signal is the sum of several components associated to different spreading codes, also leading to signals with high envelope fluctuations and high PMEPR. To overcome the strong linearity requirements for the power amplifiers in these situations, it is desirable to reduce the envelope fluctuations of the transmitted signal.

In this chapter we consider the use of nonlinear transmitters to reduce the envelope fluctuations in both DS-CDMA and MC-CDMA signals. We propose enhanced receiver structures that combine signal detection with iterative estimation and cancelation of deliberate nonlinear distortion effects introduced at the transmitter.

The publications resulting from the work presented in this chapter are [31, 32, 70–80].

The organization of this chapter is the following: after the introductory section (Section 5.1), Section 5.2 is dedicated to the transmitter and receiver structures for the downlink transmission of low-PMEPR DS-CDMA signals. This section presents a statistical characterization of the transmitted signals, which is used for performance evaluation purposes, the derivation of the receiver parameters and the derivation the nonlinear distortion estimates used to compensate the deliberate nonlinear effects introduced at the transmitter. Some aspects on implementation issues and a set of performance results and corresponding discussion ends this section. Section 5.3 presents an extension of the transmitter and receiver structures presented in Section 5.2

to the uplink transmission of low-PMEPR MC-CDMA signals, including also a detailed analysis on the receiver design and a set of performance results.

## 5.1 Introduction

As with other multicarrier schemes, MC-CDMA signals have strong envelope fluctuations and high PMEPR values, leading to amplification difficulties. Similarly, the envelope fluctuations and the PMEPR of DS-CDMA signals can be very high when we combine a large number of signals with different spreading codes, namely at the downlink transmission and/or for multicode CDMA schemes [68], making them also very prone to nonlinear effects. These nonlinear effects can be intentional, such as the ones inherent to a nonlinear signal processing for reducing the envelope fluctuations, as in [29, 30, 81, 82], or the equivalent envelope clipping characteristic that results when a nonlinear amplifier is linearized by employing suitable pre-distortion techniques [83]. They can also be inherent to a nonlinear power amplification [84, 85]. For this reason, it is desirable to reduce the envelope fluctuations of the transmitted signals. This is particularly important for the uplink transmission in MC-CDMA systems, since an efficient, low-cost power amplification is desirable at the MTs. Several techniques have been recommended for reducing the envelope fluctuations of multicarrier signals [28–31] as well as for DS-CDMA signals, e.g., by adding unused codes [69] or through suitable signal processing techniques [86–88].

A simple and promising approach for reducing the PMEPR of both MC-CDMA and DS-CDMA is to employ nonlinear clipping techniques in the time-domain, combined with a frequency-domain filtering operation so as to generate a low-PMEPR version of the signals to be transmitted, while maintaining the spectral occupation of conventional schemes [30, 87, 88] (similar techniques have also been proposed for reducing the envelope fluctuations of OFDM signals [81]). However, the filtering operation produces some regrowth on the envelope fluctuations, limiting the achievable PMEPR [88]. As with OFDM schemes [89, 90], by repeating the clipping and filtering (C&F) procedures we can further reduce the PMEPR of the transmitted signals, although the nonlinear distortion effects can be severe when a transmission with very low PMEPR values is intended [30, 32], leading to performance degradation. This performance degradation can be particularly high when we have different powers assigned to different spreading codes, especially for the spreading codes with lower power [73]. A scenario where this effect might be significant is for multi-resolution broadcasting systems [91, 92], where we transmit simultaneously several parallel data streams with different powers so as to have different error protections. This can be achieved by assigning to each resolution a subset of the available spreading codes and a different power to each subset (i.e., the spreading codes with higher power have higher error protection and, therefore, are associated to the basic (lower) resolution). Another scenario with spreading codes with different powers is when each code is intended to a given user and we assign different powers to compensate the propagation losses for each user (a kind of near-far problem created at the BS).

It was shown in [93, 94] that the use of iterative receivers with estimation and cancelation of nonlinear distortion effects can improve significantly the performance of OFDM signals with strong nonlinear distortion effects. As we will show in this chapter, these techniques can also be very effective for MC-CDMA schemes [72] and also for multicode DS-CDMA systems [73]. However, error propagation effects preclude an efficient estimation and cancelation of nonlinear distortion effects when in the presence of severe nonlinear distortion and/or at moderate and low SNR. This is particularly serious for high system load [72] and especially important when the spreading factor is small and/or if we decrease the clipping level, to reduce further the PMEPR of the transmitted signals. In fact, for those conditions the performance with iterative estimation and cancelation of nonlinear effects can be worse than without compensation [94]. When suitable channel coding schemes are employed the working region corresponds usually to low or moderate SNR values, which reduces the interest of those techniques for the estimation and cancelation of nonlinear distortion effects.

In this chapter we define iterative receivers for DS-CDMA and MC-CDMA signals with low envelope fluctuation that jointly performs the detection and the estimation and cancelation of nonlinear distortion effects that are inherent to the transmitted signals. To improve the performance at low-to-moderate SNR we consider turbo receivers combined with a threshold-based cancelation of nonlinear distortion effects. We also consider a turbo variant of our receivers where the channel decoding is performed for each iteration. We include the statistical characterization of the transmitted signals.

## 5.2 Nonlinear Effects in DS-CDMA Signals

### 5.2.1 Transmitter Structure

Let us consider the downlink transmission in DS-CDMA systems corresponding to a multi-resolution broadcasting system involving the transmission
of several parallel streams with different assigned powers so as to achieve
multi-resolution. The BS simultaneously transmits data blocks for $N_R$ resolutions. For the sake of simplicity, we assume an orthogonal spreading with
$K_r$ spreading codes associated to the $r$th resolution and the same spreading
factor $K$ for all spreading codes. This means that

$$\sum_{r=1}^{N_R} K_r \leq K.$$

(5.1)

We have a separate channel coding chain for each resolution (channel en-
coder, interleaver, etc.), as shown in Fig. 5.1(a). The coded bits associated to
the $r$th resolution are interleaved and mapped in the symbols $\{a_{m,r'}; r' \in \Psi_r\}$,
with $\Psi_r$ denoting a set with the indexes $r'$ of the spreading codes associated
to the $r$th resolution (naturally, it is assumed that $\Psi_{r_1} \cap \Psi_{r_2} = \emptyset$ for $r_1 \neq r_2$,
i.e., different spreading codes are assigned to different resolutions). For the
sake of simplicity, we assume that all $K$ orthogonal spreading codes are
used, corresponding to an equality in (5.1) (the extension to other cases is
straightforward), which means that

$$\bigcup_{r=1}^{N_R} \Psi_r = \{0, 1, \ldots, K-1\}.$$

(5.2)

The block of chips to be transmitted by the BS is $\{s_n; n = 0, 1, \ldots, N-1\}$,
where the "overall" chip symbol, $s_n$, is given by

$$s_n = \sum_{r=1}^{N_R} \sum_{r' \in \Psi_r} \xi_r s_{n,r'},$$

(5.3)

with

$$s_{n,r'} = c_{n,r'} a_{\lfloor n/K \rfloor, r'}$$

(5.4)

denoting the $n$th chip for the $r'$th spreading code, where $\{c_{n,r'}; n = 0, 1, \ldots,$
$N-1\}$ is the corresponding spreading sequence. The power assigned to the
$r$th resolution is proportional to $|\xi_r|^2$.

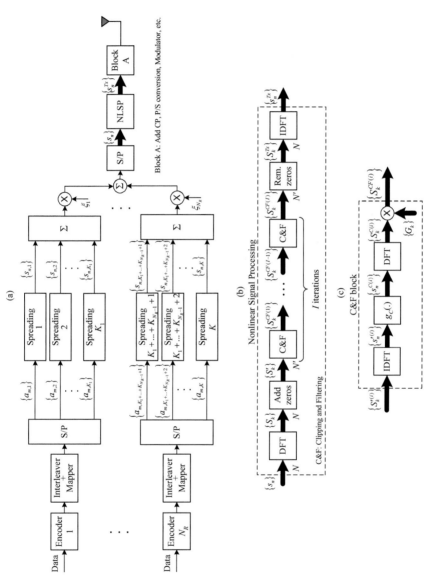

Figure 5.1  (a) DS-CDMA nonlinear transmitter structure, (b) detail of the nonlinear signal processing and (c) C&F block.

When the number of spreading codes associated to each resolutions is high (i.e., $K_r \gg 1$, especially for the spreading with higher assigned power) the coefficients $s_n$ can be regarded as samples of a zero-mean complex Gaussian process with variance given by

$$\sigma^2 = \frac{1}{2} E\left[|s_n|^2\right] = \sum_{r=1}^{N_R} \xi_r^2 \tag{5.5}$$

(again, it is assumed that $|c_{n,r'}| = 1$ and $|a_{m,r'}| = 1$). It can be shown that for normalized square-root cosine pulses with small roll-off factor the envelope distribution for the transmitted signal is identical to the distribution of $|s_n|$, i.e.,

$$p_R(R) = \frac{R}{\sigma^2} \exp\left(-\frac{R^2}{2\sigma^2}\right), \tag{5.6}$$

with $R = |s(t)|$ and

$$s(t) = \sum_{n=-\infty}^{+\infty} s_n h_T(t - nT_c), \tag{5.7}$$

where $T_c$ is the chip duration and $h_T(t)$ is a given pulse shape. Since the highest envelope values have a very small probability, the PMEPR can be defined in a statistical way as

$$\text{PMEPR} \triangleq \frac{X^2(P_x)}{2\sigma^2}, \tag{5.8}$$

where $X(P_x)$ is the envelope value that is exceeded with probability $P_x$. For a Rayleigh-distributed envelope,

$$P_x = \Pr(R > X) = \int_X^{+\infty} p_R(R)dR = \exp\left(-\frac{X^2}{2\sigma^2}\right) \tag{5.9}$$

and $X(P_x) = \sqrt{-2\sigma^2 \ln(P_x)}$. A reasonable value for $P_x$ is $10^{-3}$, which corresponds to PMEPR $\approx 8.4$ dB, regardless the value of $K_r$ (as long as $K_r \gg 1$).

The amplification of signals with high envelope fluctuations, associated with high PMEPR, is very difficult to accomplish for typical efficient low-cost amplifiers (see Appendix E), since they are required to have linear characteristics and/or a significant input back-off has to be adopted. Therefore, a reduced power efficiency precluding the necessary power levels is the price to pay for maintaining linearity and efficiency over a wide frequency band. For these reasons, it is desirable to reduce the PMEPR of the transmitted signals.

To reduce the PMEPR of the transmitted signals we consider the transmitter structure proposed in [30] and depicted in Fig. 5.1(b). The block of modified samples $\{s_n^{Tx}; n = 0, 1, \ldots, N - 1\}$ is formed from the original block of samples $\{s_n; n = 0, 1, \ldots, N - 1\}$ in the following way: The original block of samples is passed to the frequency-domain by a $N$-point DFT, leading to the block $\{S_k; k = 0, 1, \ldots, N - 1\}$. Then $N' - N$ zeros (i.e., $N' - N$ idle subcarriers) are added to the block $\{S_k; k = 0, 1 \ldots, N - 1\}$ so as to form the augmented block $\{S_k'; k = 0, 1 \ldots, N' - 1\}$, with

$$S_k' = \begin{cases} S_k, & 0 \le k < \frac{N}{2} - 1 \\ S_{k-N'}, & N' - \frac{N}{2} \le k < N' - 1 \\ 0, & \text{otherwise.} \end{cases} \tag{5.10}$$

An IDFT brings the augmented block $\{S_k'; k = 0, 1 \ldots, N' - 1\}$ back to the time-domain (see Fig. 5.1(c)), resulting the block $\{s_n'; n = 0, 1, \ldots, N' - 1\}$. These time-domain samples, which can be regarded as a sampled version of the DS-CDMA block, with the oversampling factor[1] $M_{Tx} = N'/N$, are submitted to a nonlinear operation so as to reduce the corresponding PMEPR, leading to the modified samples

$$s_n^C = g_C(|s_n'|) \exp\left(j \arg(s_n')\right) \tag{5.11}$$

(the corresponding AM-to-AM and AM-to-PM conversions are $|g_C(\cdot)|$ and $\arg(g_C(\cdot))$, respectively).

A possible nonlinear characteristic, depicted in Fig. 5.2, is an ideal envelope clipping with clipping level $s_M$, i.e.,

$$g_C(|s_n'|) = \begin{cases} |s_n'|, & |s_n'| < s_M \\ s_M, & |s_n'| \ge s_M. \end{cases} \tag{5.12}$$

A DFT brings the nonlinearly modified samples back to the frequency-domain, leading to the block $\{S_k^C; k = 0, 1, \ldots, N' - 1\}$, where a shaping operation corresponding to a frequency-domain filtering so as to reduce

---

[1] As shown in [88], $M_{Tx} > 1$ reduces the "in-band self-interference" effects of the non-linearity, while increasing the "out-of-band self-interference" levels. The oversampling is also required for an effective PMEPR reduction since the envelope excursions of the samples are only similar to the excursions of the corresponding analog signal for an oversampling factor of, at least, 2 or 4.

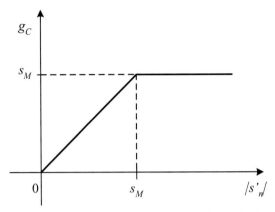

Figure 5.2 AM-to-AM conversion function of an ideal envelope clipping (zero AM-to-PM conversion).

the out-of-band radiation levels inherent to the nonlinear operation [81] is performed, leading to the block $\{S_k^{CF} = S_k^C G_k; k = 0, 1, \ldots, N' - 1\}$, with

$$G_k = \begin{cases} 1, & 0 \leq k < \frac{N}{2} - 1, \ N' - \frac{N}{2} \leq k < N' - 1 \\ 0, & \text{otherwise.} \end{cases} \qquad (5.13)$$

To reduce the PMEPR regrowth associated to the filtering operation, the signal processing operations which lead from $\{S_k'; k = 0, 1 \ldots, N' - 1\}$ to $\{S_k^{CF}; k = 0, 1 \ldots, N' - 1\}$ in Fig. 5.1(c) are repeated, in an iterative way, $I$ times:

$$S_k'^{(i)} = \begin{cases} S_k', & i = 1 \\ S_k^{CF(i-1)}, & i > 1, \end{cases} \qquad (5.14)$$

where each superscript $i$ concerns a given C&F iteration. From the block $\{S_k^{CF(I)}; k = 0, 1, \ldots, N' - 1\}$, the "final" frequency-domain block $\{S_k^{Tx}; k = 0, 1, \ldots, N - 1\}$ is formed by removing $N' - N$ zero-valued frequency-domain samples, i.e.,

$$S_k^{Tx} = \begin{cases} S_k^{CF(I)}, & 0 \leq k \leq \frac{N}{2} - 1 \\ S_{N'-N+k}^{CF(I)}, & \frac{N}{2} \leq k \leq N - 1. \end{cases} \qquad (5.15)$$

Finally, the corresponding IDFT is computed, leading to the block of modified samples $\{s_n^{Tx}; n = 0, 1, \ldots, N - 1\}$. The rest of the transmitter is similar to a conventional, CP-assisted DS-CDMA transmitter (CP insertion, D/A conversion, etc.).

For a given size-$N$ input block with duration $T$, a specific signal processing scheme can be designed through the selection of $M_{Tx} = N'/N$,
the nonlinear device and the number of C&F iterations. The block $\{s'_n; n = 0, 1, \ldots, N' - 1\}$ can be regarded as a sampled version of

$$s(t) = \sum_{n=-\infty}^{+\infty} s'_n h_T \left( t - n \frac{T}{N'} \right), \tag{5.16}$$

with the oversampling factor $M_{Tx}$, provided that the roll-off factor of the
reconstruction filter $h_T(t)$ is small. Clearly, the PMEPR of the transmitted
signal depends on the adopted pulse shape $h_T(t)$. For a square-root raised
cosine pulse there is a slight increase in the PMEPR with the roll-off factor
(less than 1 dB [88]).

The nonlinear operation can be selected so as to ensure a PMEPR reduc-
tion, and the subsequent frequency-domain operation using the set $\{G_k; k = 0, 1 \ldots, N' - 1\}$ provides a complementary filtering effect, eliminating the
out-of-band distortion effects of the nonlinearity. However, this filtering
operation produces some regrowth on the envelope fluctuations. By re-
peatedly using, in an iterative way, the nonlinear operation and the subsequent
frequency-domain filtering, we can achieve lower envelope fluctuations while
preserving a low out-of-band radiation level. For instance, Fig. 5.3 shows the
impact of the number of iterations on the histogram of $|s_n^{CF}|$ after 1, 2, 4 and
8 C&F operations for normalized clipping levels $s_M/\sigma = 0.5$, 1.0 and 2.0,
with $\sigma^2 = E[|s'_n|^2]/2$, as well as the histogram of $|s'_n|$. From this figure we
can see that with just 4 iterations the maximum envelope can be close to $s_M$.
Figures 5.4 to 5.6 show the corresponding PMEPR after 1, 2, 4 and 8 iteration
for normalized clipping levels of $s_M/\sigma = 0.5$, 1.0 and 2.0, respectively, and
Fig. 5.7 show the evolution of the PMEPR values, exceeded with probability
$P_x = 10^{-3}$, with the normalized clipping level, once again, after 1, 2, 4 and
8 iteration. Clearly, a significant reduction on the PMEPR can be achieved
with the iteration number. For instance, for a normalized clipping level of
$s_M/\sigma = 0.5$ the PMEPR can be as low as 1.7 dB after 8 C&F operations (for
conventional DS-CDMA schemes with a large number of spreading codes
PMEPR$\approx 8.4$ dB).

## 5.2.2 Statistical Characterization of the Transmitted Signal

In this subsection we present a statistical characterization of the modified
time-domain samples $\{s_n^{Tx}; n = 0, 1, \ldots, N - 1\}$ that replace the block of
time-domain samples of conventional DS-CDMA, $\{s_n; n = 0, 1, \ldots, N - 1\}$.

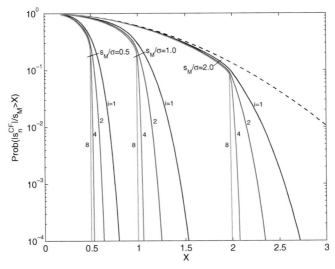

Figure 5.3 Histogram of $|s_n^{CF}|$ after 1, 2, 4 and 8 C&F operations for normalized clipping levels of $s_M/\sigma = 0.5$, 1.0 and 2.0, as well as the histogram of $|s_n'|$ (- - -).

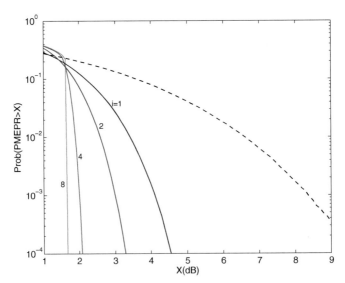

Figure 5.4 PMEPR of the transmitted signal after 1, 2, 4 and 8 C&F operations for a normalized clipping level of $s_M/\sigma = 0.5$, as well as the PMEPR of conventional DS-CDMA signal (- - -).

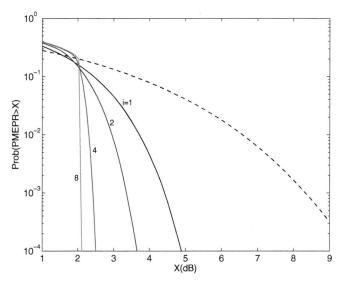

Figure 5.5  As in Fig. 5.4 but with $s_M/\sigma = 1.0$.

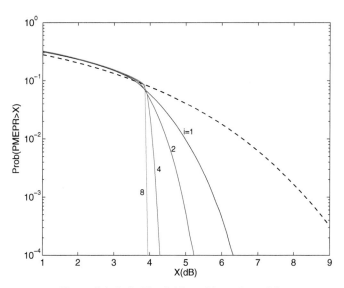

Figure 5.6  As in Fig. 5.4 but with $s_M/\sigma = 2.0$.

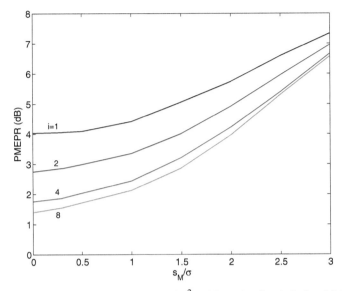

Figure 5.7 Evolution of the PMEPR ($P_x = 10^{-3}$) with $s_M/\sigma$ after 1, 2, 4 and 8 iterations.

This characterization is accurate whenever the number of spreading codes is high enough (say, several tens of spreading codes) to allow a Gaussian approximation of conventional DS-CDMA signals (as mentioned before, to validate the Gaussian approximation for $s_n$ the power associated to a given spreading code cannot be a significant fraction of the total power (see Fig. 5.8, where $N = K = 256)^2$). This statistical characterization can then be used for performance evaluation purposes, as described in the following subsection.

**Basic Signal Processing Scheme**

Let us assume that the signal at the input of the memoryless nonlinear device has a Gaussian nature. In that case, it is well-known that the signal at the nonlinearity output can be decomposed into uncorrelated "useful" and "self-interference" components [95] (see Appendix F):

$$s_n^C = \alpha s_n' + d_n, \tag{5.17}$$

---

[2] It should be pointed out that for the multi-resolution scheme we need to have several spreading codes associated to each resolution to validate a Gaussian approximation for $s_n$ (i.e., we need to have $K_r \gg 1$, especially for the spreading codes with higher assigned power).

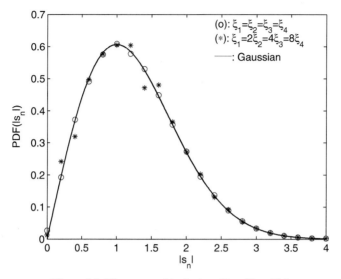

Figure 5.8  Histogram of $|s_n|$ when $N = K = 256$.

where $E[s'_n d^*_{n'}] = 0$ and

$$\alpha = \frac{E\left[s^C_n s'^*_n\right]}{E\left[\left|s'_n\right|^2\right]} = \frac{E\left[R\, g_C(R)\right]}{E\left[R^2\right]}, \tag{5.18}$$

with $R = |s'_n|$. Since $R$ has a Rayleigh distribution, we get

$$\alpha = \frac{1}{2\sigma^2} \int_0^{+\infty} R\, g_C(R) \frac{R}{\sigma^2} \exp\left(-\frac{R^2}{2\sigma^2}\right) dR. \tag{5.19}$$

When $g_C(R)$ is an ideal envelope clipping given by (5.12) then

$$\alpha = 1 - \exp\left(-\frac{s^2_M}{2\sigma^2}\right) + \frac{\sqrt{2\pi}\, s_M}{2\sigma} Q\left(\frac{s_M}{\sigma}\right). \tag{5.20}$$

Figure 5.9 shows the evolution of $\alpha$ with the the normalized clipping level $s_M/\sigma$. Clearly, the average power of the useful component at the nonlinearity output is

$$P^S_{NL} = |\alpha|^2 \sigma^2 \tag{5.21}$$

and the average power of the self-interference component is given by

$$P^I_{NL} = P_{NL} - P^S_{NL}, \tag{5.22}$$

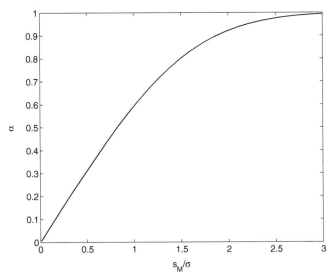

Figure 5.9 Evolution of $\alpha$ with $s_M/\sigma$.

where

$$
\begin{aligned}
P_{NL} &= \frac{1}{2} E\left[|s_n^C|^2\right] = \frac{1}{2} E\left[|g_C(R)|^2\right] \\
&= \frac{1}{2} \int_0^{+\infty} |g_C(R)|^2 \frac{R}{\sigma^2} \exp\left(-\frac{R^2}{2\sigma^2}\right) dR
\end{aligned}
\tag{5.23}
$$

denotes the average power of the signal at the nonlinearity output.
It can be shown that the autocorrelation of the output samples

$$
R_s^C(n - n') = E\left[s_n^C s_{n'}^{C*}\right]
\tag{5.24}
$$

can be expressed as a function of the autocorrelation of the input samples

$$
R_s(n - n') = E\left[s_n' s_{n'}'^{*}\right]
\tag{5.25}
$$

as follows [88] (see Appendix F):

$$
\begin{aligned}
E\left[s_n^C s_{n'}^{C*}\right] &= R_s^C(n - n') \\
&= \sum_{\gamma=0}^{+\infty} 2 P_{2\gamma+1} \frac{\left(R_s(n - n')\right)^{\gamma+1} \left(R_s^*(n - n')\right)^{\gamma}}{\left(R_s(0)\right)^{2\gamma+1}},
\end{aligned}
\tag{5.26}
$$

where the coefficient $P_{2\gamma+1}$ denotes the total power associated to the inter-modulation product (IMP) of order $2\gamma + 1$, which can be calculated as described in [88], following [96, 97] (see Appendix F). If

$$E\left[s_n s_{n'}^*\right] = \begin{cases} 2\sigma_s^2, & n = n' \\ 0, & n \neq n', \end{cases} \tag{5.27}$$

then

$$E\left[S_k S_{k'}^*\right] = \begin{cases} 2N\sigma_s^2, & k = k' \\ 0, & k \neq k' \end{cases} \tag{5.28}$$

and

$$E\left[s_n' s_{n'}'^*\right] = R_s(n - n')$$

$$= 2\sigma^2 \frac{\text{sinc}\left(\dfrac{N(n - n')}{N'}\right)}{\text{sinc}\left(\dfrac{n - n'}{N'}\right)} \exp\left(-\frac{j\pi(n - n')}{N'}\right), \tag{5.29}$$

$n, n' = 0, 1, \ldots, N' - 1$, with

$$\sigma^2 = \frac{N^2}{(N')^2}\sigma_s^2. \tag{5.30}$$

Since

$$R_s^C(n - n') = |\alpha|^2 R_s(n - n') + E[d_n d_{n'}^*], \tag{5.31}$$

it can be easily recognized that

$$P_1 = P_{NL}^S = |\alpha|^2 \sigma^2 \tag{5.32}$$

and

$$E\left[d_n d_{n'}^*\right] = R_d(n - n')$$

$$= \sum_{\gamma=1}^{+\infty} 2P_{2\gamma+1} \frac{\left(R_s(n - n')\right)^{\gamma+1}\left(R_s^*(n - n')\right)^{\gamma}}{\left(R_s(0)\right)^{2\gamma+1}}. \tag{5.33}$$

The total power of the self-interference term is

$$P_{NL}^I = \frac{1}{2}R_d(0) = \sum_{\gamma=1}^{+\infty} P_{2\gamma+1} = P_{NL} - P_{NL}^S. \tag{5.34}$$

Figure 5.10 shows the evolution of $P_{NL}$, $P_{NL}^S$ and $P_{NL}^I$ with the normalized clipping level $s_M/\sigma$.

Having in mind (5.17) and the signal processing chain in Fig. 5.1(c), the frequency-domain block $\{S_k^{CF} = S_k^C G_k; k = 0, 1, \ldots, N'-1\}$ can obviously be decomposed into useful and self-interference components:

$$S_k^{CF} = \alpha S_k' G_k + D_k G_k, \tag{5.35}$$

where $\{D_k; k = 0, 1, \ldots, N'-1\}$ denotes the DFT of $\{d_n; n = 0, 1, \ldots, N'-1\}$.

It can be shown that $E[D_k] = 0$ and

$$E\left[D_k D_{k'}^*\right] = \begin{cases} N' G_d(k), & k = k' \\ 0, & k \neq k', \end{cases} \tag{5.36}$$

$k, k' = 0, 1, \ldots, N'-1$, where $\{G_d(k); k = 0, 1, \ldots, N'-1\}$ is the DFT of $\{R_d(n); n = 0, 1, \ldots, N'-1\}$. Similarly,

$$E\left[S_k^C S_{k'}^{C*}\right] = \begin{cases} N' G_s^C(k), & k = k' \\ 0, & k \neq k', \end{cases} \tag{5.37}$$

where $\{G_s^C(k) = |\alpha|^2 G_s(k) + G_d(k); k = 0, 1, \ldots, N'-1\}$ is the DFT of $\{R_s^C(n); k = 0, 1, \ldots, N'-1\}$, with $R_s^C(n)$ given by (5.26), and $\{G_s(k); k = 0, 1, \ldots, N'-1\}$ the DFT of $\{R_s(n); n = 0, 1, \ldots, N'-1\}$. Therefore,

$$E\left[S_k^{CF} S_{k'}^{CF*}\right] = G_k G_{k'}^* E\left[S_k^C S_{k'}^{C*}\right] = \begin{cases} N' |G_k|^2 G_s^C(k), & k = k' \\ 0, & k \neq k'. \end{cases} \tag{5.38}$$

The "final" frequency-domain block can also be decomposed into uncorrelated useful and self-interference terms:

$$S_k^{Tx} = \alpha S_k + D_k^{Tx}, \tag{5.39}$$

with

$$D_k^{Tx} = \begin{cases} D_k, & 0 \leq k \leq \frac{N}{2} - 1 \\ D_{N'-N+k}, & \frac{N}{2} \leq k \leq N-1. \end{cases} \tag{5.40}$$

This means that

$$s_n^{Tx} = \alpha s_n + d_n^{Tx}, \tag{5.41}$$

where $\{d_n^{Tx}; n = 0, 1, \ldots, N-1\}$ is the IDFT of $\{D_k^{Tx}; k = 0, 1, \ldots, N-1\}$.

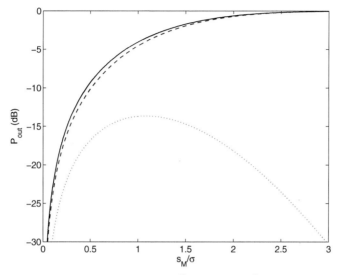

Figure 5.10 Evolution of $P_{NL}$ (——), $P_{NL}^S$ (- - -) and $P_{NL}^I$ ($\cdots$) with $s_M/\sigma$.

## Iterative Signal Processing Scheme

For the iterative signal processing scheme in Fig. 5.1(b) the Gaussian approximation for the samples at the input to the nonlinearity is no longer valid after the first iteration. Therefore, the method for modeling the transmitted blocks needs to be modified. Our simulations have shown that

$$S_k^{CF(i)} \approx \alpha_k^{(i)} S_k' G_k + D_k^{(i)} G_k, \qquad (5.42)$$

with $\alpha_k^{(i)}$ depending on $k$ when $i > 1$, as shown in Fig. 5.11. This means that the $k$th component of the frequency-domain block, for the $i$th iteration, can still be decomposed as a sum of two uncorrelated components (a similar behavior was observed in [90]). The statistical characterization concerning the first iteration, described above, can be regarded as a special case of (5.42) with constant $\alpha_k^{(1)}$; in this case, the values of $\alpha_k^{(1)}$ and $E[|D_k^{(1)}|^2]$ can be obtained analytically as described in the previous subsection. For the remaining iterations ($i > 1$) the values of $\alpha_k^{(i)}$ and $E[|D_k^{(i)}|^2]$ can be obtained by simulation in the following way:

$$\alpha_k^{(i)} = \frac{E\left[S_k^{C(i)} S_k'^*\right]}{E\left[\left|S_k'\right|^2\right]} \qquad (5.43)$$

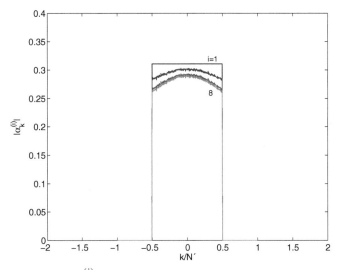

Figure 5.11 Evolution of $|\alpha_k^{(i)}|$ after 1, 2, 4 and 8 C&F operations, with a normalized clipping level of $s_M/\sigma = 0.5$.

and

$$E\left[|D_k^{(i)}|^2\right] = E\left[\left|S_k^{C(i)} - \alpha_k^{(i)} S_k'\right|^2\right],\tag{5.44}$$

respectively. Figures 5.11 and 5.12 shows the values of $\alpha_k^{(i)}$ and $E[|D_k^{(i)}|^2]$, respectively, after 1, 2, 4 and 8 iterations for a normalized clipping level of $s_M/\sigma = 0.5$. From Fig. 5.11 we can observe that when $i > 1$, $\alpha_k^{(i)} < \alpha_k^{(1)}$ for all values of $k$ and $\alpha_k^{(i)}$ is no longer constant. We can also observe from Fig. 5.12 that the self-interference levels decrease in the out-of-band region and increase in the in-band region with the iteration number. Our simulations also indicate that $E\left[D_k^{(i)}\right] = 0$ and $E\left[D_k^{(i)} D_{k'}^{(i)*}\right] \approx 0, k \neq k'$, as with the basic transmitter structure (i.e., with a single C&F procedure). From (5.15) and (5.42), it is clear that the samples $\{S_k^{Tx}; k = 0, 1, \ldots, N - 1\}$ can be decomposed into uncorrelated "useful" and "self-interference" terms:

$$S_k^{Tx} = \alpha_k^{Tx} S_k + D_k^{Tx},\tag{5.45}$$

with

$$\alpha_k^{Tx} = \begin{cases} \alpha_k^{(I)}, & 0 \leq k \leq \frac{N}{2} - 1 \\ \alpha_{N'-N+k}^{(I)}, & \frac{N}{2} \leq k \leq N - 1 \end{cases}\tag{5.46}$$

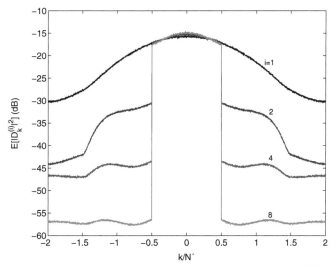

Figure 5.12 Evolution of the PSD of the self-interference component, $E[|D_k^{(i)}|^2]$, after 1, 2, 4 and 8 C&F operations, with a normalized clipping level of $s_M/\sigma = 0.5$.

and $\{D_k^{Tx}; k = 0, 1, \ldots, N - 1\}$ related to $\{D_k^{(I)}; k = 0, 1, \ldots, N' - 1\}$ as in (5.40). This means that

$$s_n^{Tx} = s_n^U + d_n^{Tx}, \tag{5.47}$$

where $\{s_n^U; n = 0, 1, \ldots, N - 1\}$ is the IDFT of $\{\alpha_k^{Tx} S_k; k = 0, 1, \ldots, N - 1\}$ (corresponding to the "useful" component of the transmitted signal) and $\{d_n^{Tx}; n = 0, 1, \ldots, N - 1\}$ is the IDFT of $\{D_k^{Tx}; k = 0, 1, \ldots, N - 1\}$ (corresponding to the "self-interference" component on the transmitted signal). Therefore, the modified samples $\{s_n^{Tx}; n = 0, 1, \ldots, N - 1\}$ can always be decomposed into uncorrelated "useful" and "self-interference" components regardless of the number of C&F procedures. However, after the first iteration, the "useful" component is no longer proportional to the original samples, $\{s_n; n = 0, 1, \ldots, N\}$, due to the filtering effect inherent to $\alpha_k^{Tx}$ (see Fig. 5.11).

## 5.2.3 Performance Evaluation

Since $E[D_k^{(l)} D_{k'}^{(l)*}] \approx 0$, $k \neq k'$, we have $E[S_k^{CF(l)} S_{k'}^{CF(l)*}] \approx 0, k \neq k'$, leading to $E[S_k^{Tx} S_{k'}^{Tx*}] \approx 0$ for $k \neq k'$. Consequently, it can be shown that the PSD of the transmitted signals is [88]

$$G_{Tx}(f) = \frac{|H_T(f)|^2}{T^3} \sum_{k=-\infty}^{+\infty} E\left[|S_k^{Tx}|^2\right] \left|R\left(f - \frac{k}{T}\right)\right|^2, \qquad (5.48)$$

with the block duration $T = NT_c$ and $R(f) = T\text{sinc}(fT)$ (it is assumed that $S_k^{Tx} = S_{k+N}^{Tx}, \forall_k$, i.e., $S_k^{Tx}$ is periodic, with period $N$). This means that the spectral occupations of the modified signal and the corresponding conventional DS-CDMA signal are similar when $G_k$ follows (5.13), regardless of the clipping level and the number of iterations.

For an ideal Gaussian channel, the detection of the $m$th symbol transmitted by the $r'$th spreading code is based on

$$\tilde{a}_{m,r'} = \sum_{n=mK}^{mK+K-1} y_n c_{n,r'}^* = \sum_{n=mK}^{mK+K-1} s_n^U c_{n,r'}^* + d_m^{eq} + w_m^{eq}, \qquad (5.49)$$

$m = 0, 1, \ldots, M - 1$, where

$$y_n = s_n^{Tx} + w_n = s_n^U + d_n^{Tx} + w_n \qquad (5.50)$$

denotes the output of the detection filter (assumed matched to the transmission filter $h_T(t)$) associated to the $n$th chip, with $w_n$ denoting the corresponding channel noise. In (5.49),

$$d_m^{eq} = \sum_{n=mK}^{mK+K-1} d_n^{Tx} c_{n,r'}^* \qquad (5.51)$$

and

$$w_m^{eq} = \sum_{n=mK}^{mK+K-1} w_n c_{n,r'}^* \qquad (5.52)$$

denote the equivalent self-interference and noise terms, respectively, for detection purposes.

Clearly, the power of the self-interference term, $d_m^{eq}$, is independent of $m$ and given by

$$P_{Tx}^I = \sum_{n=mK}^{mK+K-1} E\left[|d_n^{Tx}|^2\right] = \frac{K}{N} \sum_{n=0}^{N-1} E\left[|d_n^{Tx}|^2\right] = \frac{K}{N^2} \sum_{k=0}^{N-1} E\left[|D_k^{Tx}|^2\right].$$

$$(5.53)$$

The "useful" component for the detection of the $m$th symbol transmitted by the $r'$th spreading code is based on

$$\sum_{n=mK}^{mK+K-1} s_n^U c_{n,r'}^* \overset{(a)}{\approx} \bar{\alpha} \sum_{n=mK}^{mK+K-1} s_n c_{n,r'}^* \overset{(b)}{=} K\bar{\alpha}a_{m,r'}, \qquad (5.54)$$

where

$$\bar{\alpha} = \frac{1}{N} \sum_{k=0}^{N-1} \alpha_k^{Tx}. \qquad (5.55)$$

In (5.54), $(a)$ follows from $\alpha_k^{Tx} \approx \bar{\alpha}$ (see Fig. 5.11) and $(b)$ follows from the orthogonality of the spreading codes.

By assuming $E[|a_{m,r'}|^2] = 1$, the power of the "useful" component for the detection of the $r'$th spreading code of the $r$th resolution ($r' \in \Psi_r$) is

$$P_{Tx,r'}^S \approx |K\bar{\alpha}|^2 = \frac{|K\bar{\alpha}\xi_r|^2 \sum_{n=0}^{N-1} E\left[|s_n|^2\right]}{N \sum_{r''=1}^{N_R} K_{r''}|\xi_{r''}|^2} = \frac{K|\bar{\alpha}\xi_r|^2}{N|\overline{\xi_r}|^2} \sum_{n=0}^{N-1} E\left[|s_n|^2\right]$$

$$= \frac{K|\bar{\alpha}|^2 \eta_{\xi,r}}{N} \sum_{n=0}^{N-1} E\left[|s_n|^2\right] = \frac{K|\bar{\alpha}|^2 \eta_{\xi,r}}{N^2} \sum_{k=0}^{N-1} E\left[|S_k|^2\right]$$

$$\overset{(a)}{\approx} \frac{K\eta_{\xi,r}}{N^2} \sum_{k=0}^{N-1} E\left[|\alpha_k^{Tx} S_k|^2\right], \qquad (5.56)$$

with $(a)$ following, once again, from $\alpha_k^{Tx} \approx \bar{\alpha}$. In (5.56),

$$\overline{|\xi_r|^2} = \frac{1}{K} \sum_{r''=1}^{N_R} K_{r''}|\xi_{r''}|^2 \qquad (5.57)$$

and

$$\eta_{\xi,r} = \frac{|\xi_r|^2}{\overline{|\xi_r|^2}}. \qquad (5.58)$$

Therefore, the signal-to-self-interference ratio for the detection of the $r'$th spreading code is

$$\text{SIR}_{Tx,r'} = \frac{P_{Tx,r'}^S}{P_{Tx}^I} \approx \eta_{\xi,r'} \text{SIR}_{Tx}, \qquad (5.59)$$

where the signal-to-self-interference ratio of the transmitted signal is given by

$$\text{SIR}_{Tx} = \frac{\sum_{n=mK}^{mK+K-1} E\left[\left|s_n^U\right|^2\right]}{\sum_{n=mK}^{mK+K-1} E\left[\left|d_n^{Tx}\right|^2\right]} = \frac{\sum_{k=0}^{N-1} E\left[\left|\alpha_k^{Tx} S_k\right|^2\right]}{\sum_{k=0}^{N-1} E\left[\left|D_k^{Tx}\right|^2\right]}. \tag{5.60}$$

Clearly, the spreading codes with smaller assigned power have worse signal-to-self-interference ratio. As shown in Figs. 5.13 and 5.14, when we increase the number of iterations, the higher in-band self-interference levels combined with the corresponding decrease of $\alpha_k^{Tx}$, lead to lower $\text{SIR}_{Tx}$.

Clearly, the performance of the proposed transmitter structure is worse than the performance of a conventional DS-CDMA transmitter. On the one hand, just a fraction

$$\eta_S = \frac{\sum_{n=0}^{N-1} E\left[\left|s_n^U\right|^2\right]}{\sum_{n=0}^{N-1} E\left[\left|s_n^U\right|^2\right] + \sum_{n=0}^{N-1} E\left[\left|d_n^{Tx}\right|^2\right]} \approx \frac{\text{SIR}_{Tx}}{\text{SIR}_{Tx} + 1}, \tag{5.61}$$

of the total transmitted power is useful; on the other hand, the self-interference component $d_n^{eq}$ is added to the Gaussian channel noise.

For an iterative transmitter, an additional degradation results from the filtering effect inherent to the coefficients $\alpha_k^{Tx}$, which eliminates the orthogonality between different spreading codes, even for an ideal AWGN channel. This filtering effect could be compensated at the transmitter by multiplying $S_k^{Tx}$ by $1/\alpha_k^{Tx}$. However, this produces an additional PMEPR regrowth of a few tenths of dB. As an alternative, we could restore the orthogonality between spreading codes through an appropriate chip-level equalization at the receiver. This is especially recommendable for cyclic-prefix-aided DS-CDMA schemes employing FDE techniques [60, 98] since the equalizer can be designed for compensating both the filtering effects inherent to the coefficients $\alpha_k^{Tx}$ and the filtering effects associated to a time-dispersive channel. For an ideal AWGN channel and an FDE characterized by the multiplying coefficients $1/\alpha_k^{Tx}$, there is a performance degradation factor of

$$\eta_{MF} \approx \frac{1}{N^2} \sum_{k=0}^{N-1} E\left[\left|\alpha_k^{Tx}\right|^2\right] \sum_{k=0}^{N-1} E\left[\left|\alpha_k^{Tx}\right|^{-2}\right], \tag{5.62}$$

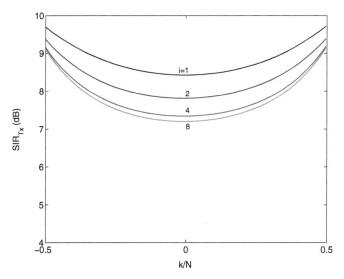

Figure 5.13 In-band SIR levels with $s_M/\sigma = 0.5$ after 1, 2, 4 and 8 iterations.

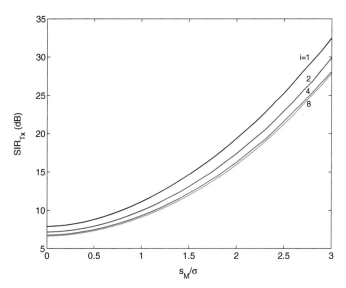

Figure 5.14 Evolution of SIR levels with $s_M/\sigma$ after 1, 2, 4 and 8 iterations.

relatively to the case where $\alpha_k^{Tx}$ is constant. Since, in our case $\alpha_k^{Tx}$ is almost constant, $\eta_{MF}$ is very close to one. The receiver described in the following section performs this compensation together with the FDE procedure.

When $K \gg 1$, the term $d_n^{eq}$ is approximately Gaussian-distributed. Therefore, if the data symbols are selected from a QPSK constellation under a Gray mapping rule (the extension to other constellations is straightforward) the BER for an ideal AWGN channel is approximately given by

$$P_b \approx Q\left(\sqrt{\text{SNR}_{Tx,r'}}\right), \tag{5.63}$$

where $\text{SNR}_{Tx,r'}$ denotes an equivalent signal-to-noise ratio for the detection of the $r'$th spreading code given by

$$\text{SNR}_{Tx,r'} = \frac{P_{Tx,r'}^S}{P_{Tx}^I + P_N^{eq}}, \tag{5.64}$$

with $P_N^{eq} = E[|w_n^{eq}|^2] = KE[|w_n|^2]$. This formula is very accurate, as shown in Fig. 5.15, where $N = K = 256$ and the normalized clipping level $s_M/\sigma = 0.5$.

It should be noted that, even if we were able to remove entirely the negative impact of the self-interference term in each frequency, we could not avoid a certain performance degradation, expressed by $\eta_S$, due to the useless transmitted power inherent to the self-interference component. Figure 5.16 shows the evolution of $\eta_S$ when $g_C(R)$ corresponds to an ideal envelope clipping, with the normalized clipping level $s_M/\sigma$, for several C&F iterations. As we can see, the values of $\eta_S$ are very close to 1 (0 dB), unless the clipping levels is very low.

## Special Cases and Simplified Formulas

For most cases of interest, the analytical approach for obtaining the SIR levels and the BER performances described above can be simplified with only a very slight decrease in its accuracy.

If there is no oversampling before the nonlinear operation (i.e., for $M_{Tx} = N'/N = 1$), then

$$P_{Tx,IB}^I = P_{Tx}^I = P_{NL}^I \tag{5.65}$$

with $P_{Tx,IB}^I$ denoting the total in-band self-interference power, and

$$\frac{P_{Tx}^S}{P_{Tx}^I} = \text{SIR}_{Tx} = \text{SIR}_{NL} \tag{5.66}$$

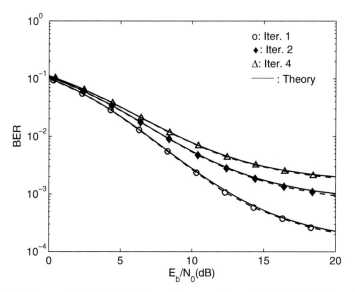

Figure 5.15 Theoretical and simulation performance for an ideal AWGN channel for $N = K = 256$ and a normalized clipping level of $s_M/\sigma = 0.5$.

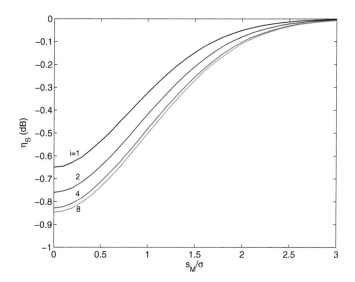

Figure 5.16 Evolution of $\eta_S$ with the normalized clipping level $s_M/\sigma$, for several C&F iterations.

where $\mathrm{SIR}_{NL}$ denotes the SIR at the output of the nonlinear device. Therefore,

$$\eta_S = \frac{\mathrm{SIR}_{NL}}{\mathrm{SIR}_{NL} + 1} \tag{5.67}$$

Let us assume now that $M_{Tx} > 1$. To obtain an approximate formula for the SIR levels that does not require the computation of all IMPs, we will assume that the total self-interference power is associated to the IMP of order 3, i.e.,

$$P_3 = P_{NL}^I \tag{5.68}$$

and

$$P_{2\gamma+1} = 0, \quad \gamma > 1 \tag{5.69}$$

(this is a slightly pessimistic assumption relatively to the in-band self-interference levels). In that case, we can derive a closed formula for the power distribution of the nonlinear self-interference component, as well as the total in-band self-interference power, as explained in the following.

Let us assume an infinite oversampling factor ($M_{Tx} = +\infty$). If all frequency-domain samples have the same average power and its number is very high then the autocorrelation of the complex envelope of the signal at the input of the nonlinear device, referred to the central frequency of the spectrum, is approximately given by

$$R_{in}(\tau) = P_{in}\mathrm{sinc}(\tau B), \tag{5.70}$$

with $P_{in}$ denoting its power and $B$ its bandwidth. The corresponding PSD is

$$G_{in}(f) = \frac{P_{in}}{B}\mathrm{rect}\left(\frac{f}{B}\right). \tag{5.71}$$

Having in mind (5.68) and (5.69), then the autocorrelation of the self-interference component is (see (5.33))

$$R_d(\tau) = 2P_3 = 2P_{NL}^I \frac{R_{in}^3(\tau)}{R_{in}^3(0)}, \tag{5.72}$$

and the corresponding PSD is

$$\begin{aligned} G_d(f) &= \mathcal{F}\{R_d(\tau)\} = 2P_{NL}^I \frac{G_{in}(f) * G_{in}(f) * G_{in}(f)}{R_{in}^3(0)} \\ &= \frac{2P_{NL}^I G_3\left(\dfrac{f}{B}\right)}{B} \end{aligned} \tag{5.73}$$

where $G_3(f) = G_1(f) * G_1(f) * G_1(f)$, with $G_1(f) = \text{rect}(f)$. Clearly,

$$
G_3(f) = \begin{cases}
\dfrac{3}{4} - f^2, & |f| \leq \dfrac{1}{2} \\[2mm]
\dfrac{9}{8} - \dfrac{3}{2}|f| + \dfrac{1}{2}f^2, & \dfrac{1}{2} < |f| < \dfrac{3}{2} \\[2mm]
0, & |f| \geq \dfrac{3}{2}.
\end{cases} \tag{5.74}
$$

Therefore, the total in-band power is

$$
P_{NL,IB} = \int_{-\frac{B}{2}}^{\frac{B}{2}} \left( G_d(f) + G_d(f - M_{T_x} B) + G_d(f + M_{T_x} B) \right) df. \tag{5.75}
$$

It can be shown that, in this case,

$$
P_{NL,IB} = \kappa(M_{Tx}) P_{NL}^I, \tag{5.76}
$$

with

$$
\kappa(M_{Tx}) = \begin{cases}
\dfrac{1}{3}\left(-M_{Tx}^3 + 6M_{Tx}^2 - 12M_{Tx} + 10\right), & 1 \leq M_{Tx} < 2 \\[2mm]
\dfrac{2}{3}, & M_{Tx} \geq 2.
\end{cases} \tag{5.77}
$$

Using these formulas, the average power of the self-interference component for the $N$ in-band frequency-domain samples is

$$
\frac{1}{N} \sum_{k=0}^{N-1} E\left[\left|D_k^{Tx}\right|^2\right] \approx \frac{\kappa(M_{Tx})}{N} \sum_{k=0}^{N-1} E\left[\left|D_k^{Tx}\right|^2\right]
$$
$$
= \frac{\kappa(M_{Tx})}{N} 2P_{NL}^I. \tag{5.78}
$$

This means that, for $G_k$ given by (5.13),

$$
\eta_S \approx \frac{\dfrac{\text{SIR}_{NL}}{\kappa(M_{Tx})}}{\dfrac{\text{SIR}_{NL}}{\kappa(M_{Tx})} + 1}. \tag{5.79}
$$

For $M_{Tx} \geq 2$, $\kappa(M_{Tx}) = 2/3$ and we have a gain of $3/2$ (i.e., approximately 1.8 dB) in the equivalent SIR levels relatively to the case where there is

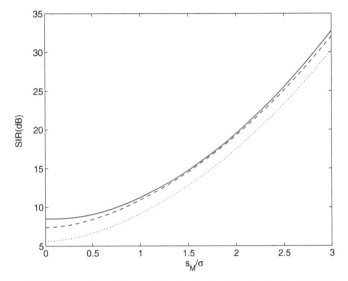

Figure 5.17  $\text{SIR}_{Tx}$ when $M_{Tx} = 1$ ($\cdots$) or 2 (-----) and given by (5.80) (- - -).

no oversampling ($M_{Tx} = 1$). Figure 5.17 shows the impact of the normalized clipping level on $\text{SIR}_{Tx}$ when $M_{Tx} = 1$ or 2. We also include the approximate $\text{SIR}_{Tx}$ formula that is obtained by using (5.76), with $M_{Tx} \geq 2$, i.e.,

$$\text{SIR}_{Tx} \approx \frac{3}{2}\text{SIR}_{NL}. \tag{5.80}$$

From this figure, it is clear that approximation (5.80) is very accurate, especially for moderate clipping levels.

The computation of $\eta_S$ involves only two integrals inherent to $\alpha$ and $P_{NL}$ (see (5.19) and (5.23)). If the nonlinearity corresponds to an ideal envelope clipping, i.e., when $g_C(R)$ is given by (5.12), which is a very common situation, these two integrals can be written in a closed form as (5.20) and

$$P_{NL} = \sigma^2 \left(1 - \exp\left(-\frac{s_M^2}{2\sigma^2}\right)\right). \tag{5.81}$$

## 5.2.4 Receiver Design

### Receiver Structure

Since the orthogonality between spreading codes is lost in a time-dispersive channel, we perform an FDE before the "despreading" procedure [48,

60]: after removing the CP, the received time-domain block $\{y_n; n = 0, 1, \ldots, N-1\}$ is passed to the frequency-domain by a DFT leading to the block $\{Y_k; k = 0, 1, \ldots, N-1\}$, where

$$Y_k = H_k S_k^{Tx} + N_k. \tag{5.82}$$

Considering first a linear receiver, the optimum FDE coefficients, in the MMSE sense, can be obtained by employing the Lagrange multipliers' method, as described in Section 3.1.2. Since

$$S_k^{Tx} = \alpha_k^{Tx} S_k + D_k^{Tx}, \tag{5.83}$$

it is easy to show that

$$F_k = \frac{\mathcal{K}_F \alpha_k^{Tx*} E\left[|S_k|^2\right] H_k^*}{\left(E\left[|\alpha_k^{Tx} S_k|^2\right] + E\left[|D_k^{Tx}|^2\right]\right)|H_k|^2 + E\left[|N_k|^2\right]}, \tag{5.84}$$

where $\mathcal{K}_F$ is given by (3.21), i.e.,

$$\mathcal{K}_F = 1 - \frac{\lambda}{2\sigma_S^2 N} \tag{5.85}$$

ensures that $\sum_{k=0}^{N-1} F_k H_k \alpha_k^{Tx*}/N = 1$ and $E[|D_k^{Tx}|^2]$ can be obtained as described in Section 5.2.2. The frequency-domain block at the output of the FDE is then $\{\tilde{S}_k; k = 0, 1, \ldots, N-1\}$, with

$$\tilde{S}_k = Y_k F_k. \tag{5.86}$$

The data block associated with the $r'$th spreading codes could be estimated by despreading the time-domain block at the FDE output, $\{\tilde{s}_n; n = 0, 1, \ldots, N-1\}$, i.e., from the samples

$$\tilde{a}_{m,r'} = \sum_{n'=nK}^{nK+K} \tilde{s}_{n'} c_{n',r'}^*, \tag{5.87}$$

where $\{\tilde{s}_n; n = 0, 1, \ldots, N-1\}$ is the IDFT of $\{\tilde{S}_k; k = 0, 1, \ldots, N-1\}$).

As expected, the nonlinear effects lead to some BER degradation relatively to conventional DS-CDMA schemes, especially when a low PMEPR is intended and/or for codes with small assigned power. This degradation results from both the "useless" transmitted power spent on self-interference and the received self-interference being added to the channel noise.

To improve performance we consider the iterative receiver for CP-assisted multicode DS-CDMA depicted in Fig. 5.18.

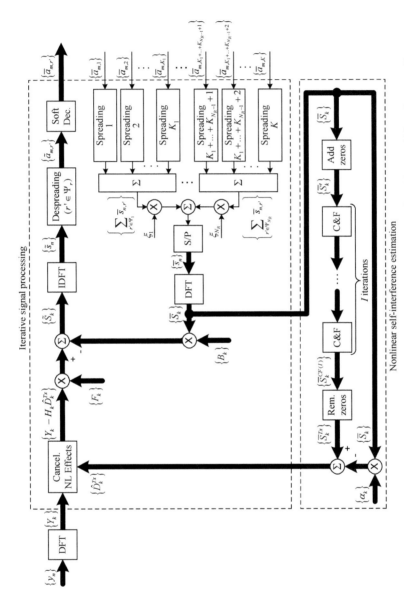

Figure 5.18  Iterative receiver structure with estimation and compensation of nonlinear self-interference effects.

This receiver employs an IB-DFE with soft decisions [99, 100] (instead of the linear FDE) which is combined with estimation and compensation of nonlinear self-interference components. The receiver can be described as follows: For a given iteration, the signal at the output of the FDE is

$$\tilde{S}_k = F_k \left( Y_k - H_k \hat{D}_k^{Tx} \right) - B_k \overline{S}_k, \tag{5.88}$$

where $\{B_k; k = 0, 1, \ldots, N - 1\}$ is the block of feedback coefficients, $\{\overline{S}_k; k = 0, 1, \ldots, N-1\}$ are the average values of $\{S_k; k = 0, 1, \ldots, N-1\}$ associated to the previous iteration, conditioned to the FDE output, and $\{\hat{D}_k^{Tx}; k = 0, 1, \ldots, N - 1\}$ are estimates of the transmitted nonlinear self-interference components obtained by submitting $\overline{S}_k$ to a replica of nonlinear device at the transmitter (see Fig. 5.18) (it is assumed that the nonlinear characteristic adopted at the transmitter, $g_C(R)$, is known at the receiver). After the despreading operation, the data estimates for each spreading code $\{\overline{a}_{m,r'}; m = 0, 1, \ldots, M - 1\}$ are obtained by submitting $\{\tilde{a}_{m,r'}; m = 0, 1, \ldots, M - 1\}$ to a soft-decision device and used to form the estimates of the chip samples $\{\overline{s}_n; n = 0, 1, \ldots, N - 1\}$.

As with the turbo receivers for linear transmitters described in Section 3.2.1, we can also define a receiver structure, that will be denoted as Turbo FDE, with SISO channel decoder outputs employed in the feedback loop.

### Derivation of the Receiver Parameters
Once again, by employing the Lagrange multipliers' method (see Section 3.1.2), the optimum values of the feedforward and feedback coefficients, in the MMSE sense, are

$$F_k = \frac{\mathcal{K}_F \alpha_k^{Tx*} H_k^*}{\beta + \eta_k |H_k|^2 + \left(1 - \rho^2\right) \left|\alpha_k^{Tx} H_k\right|^2} \tag{5.89}$$

and

$$B_k = F_k H_k \alpha_k^{Tx} - 1, \tag{5.90}$$

respectively, with

$$\eta_k = \frac{E\left[\left|D_k^{Res}\right|^2\right]}{E\left[|S_k|^2\right]}, \tag{5.91}$$

where $D_k^{Res} = D_k^{Tx} - \hat{D}_k^{Tx}$ denotes the residual nonlinear self-interference, and the correlation coefficient $\rho$ given by

$$\rho = \frac{\sum_{r=1}^{N_R} \sum_{r' \in \Psi_r} \xi_r^2 \rho_{r'}}{\sum_{r=1}^{N_R} \xi_r^2 K_r}, \tag{5.92}$$

corresponding to a weighted average of the different $\rho_{r'}, r' \in \Psi_r$, with

$$\rho_{r'} = \frac{1}{2M} \sum_{m=0}^{M-1} \left( \rho_{m,r'}^I + \rho_{m,r'}^Q \right) \tag{5.93}$$

(as in (3.49)).

Note that, since $\rho = 0$ for the first iteration, the receiver reduces to a linear FDE with feedforward coefficients given by (5.84). After the first iteration, and if the residual BER is not too high (at least for the spreading codes for which a higher transmit power is associated), we have $\overline{a}_{m,r'} \approx a_{m,r'}$ for most of the data symbols, leading to $\overline{S}_k \approx S_k$. Consequently, we can use the feedback coefficients to eliminate a significant part of the residual interference.

The "overall chip averages" are given by

$$\overline{s}_n = \sum_{r=1}^{N_R} \xi_r \sum_{r' \in \Psi_r} c_{n,r'} \overline{a}_{\lfloor n/K \rfloor, r'} \tag{5.94}$$

and, as in (4.45)–(4.48), the average data values are

$$\overline{a}_{m,r'} = \tanh \left( \frac{\tilde{a}_{m,r'}^I}{\sigma_{r'}^2} \right) + \tanh \left( \frac{\tilde{a}_{m,r'}^Q}{\sigma_{r'}^2} \right), \tag{5.95}$$

where $\tilde{a}_{m,r'} = \tilde{a}_{m,r'}^I + j\tilde{a}_{m,r'}^Q$ denotes the despreaded symbols and

$$\sigma_{r'}^2 = \frac{1}{2} E \left[ |a_{m,r'} - \tilde{a}_{m,r'}|^2 \right] \approx \frac{1}{2M} \sum_{m=0}^{M-1} E \left[ |\hat{a}_{m,r'} - \tilde{a}_{m,r'}|^2 \right], \tag{5.96}$$

with $\hat{a}_{m,r'}$ denoting the hard decisions associated to $\tilde{a}_{m,r'}$.

**Derivation of the Nonlinear Distortion Estimates**

An important issue in our receiver is the estimation of nonlinear distortion effects. With respect to $\eta_k$, for the first iteration $\hat{D}_k^{Tx} = 0$ and we can use the method described in Section 5.2.2 to obtain $E[|D_k^{Tx}|^2]$. For the remaining iterations it has to be obtained by simulation. One possibility is to submit the soft-decision chip estimates $\{\overline{s}_n; n = 0, 1, \ldots, N - 1\}$ to a replica of the nonlinear signal processing chain at the transmitter so as to obtain the self-interference estimates

$$\hat{D}_k^{Tx}\Big|_{\{\overline{s}_n\}} = \overline{S}_k^{Tx}\Big|_{\{\overline{s}_n\}} - \alpha_k^{Tx}\overline{S}_k, \tag{5.97}$$

where $\{\overline{S}_k; k = 0, 1, \ldots, N - 1\}$ is the DFT of $\{\overline{s}_n; n = 0, 1, \ldots, N - 1\}$ (as shown in Fig. 5.18).

As an alternative, we could also obtain the nonlinear distortion estimates by submitting the hard decisions $\{\hat{s}_n; n = 0, 1, \ldots, N - 1\}$ to the nonlinear signal processing chain instead of $\{\overline{s}_n; n = 0, 1, \ldots, N - 1\}$, i.e.,

$$\hat{D}_k^{Tx}\Big|_{\{\hat{s}_n\}} = \hat{S}_k^{Tx}\Big|_{\{\hat{s}_n\}} - \alpha_k^{Tx}\hat{S}_k, \tag{5.98}$$

where $\{\hat{S}_k; k = 0, 1, \ldots, N - 1\}$ is the DFT of $\{\hat{s}_n; n = 0, 1, \ldots, N - 1\}$. To avoid error propagation, these estimates could be weighted by the correlation coefficient $\rho$, which can be regarded as the overall reliability of the decisions used in the feedback loop. This leads to the estimates

$$\hat{D}_k^{Tx} = \rho \, \hat{D}_k^{Tx}\Big|_{\{\hat{s}_n\}}. \tag{5.99}$$

Figures 5.19 to 5.21 show the behavior of $E[|D_k^{Res}|^2]$ as a function of $\rho$ for the three estimates described above, for normalized clipping levels of $s_M/\sigma = 0.5$, 1.0 and 2.0, respectively, (similar behavior was observed for other values of $s_M/\sigma$). From these figures we can observe that, for small values of $\rho$, using the estimates to remove nonlinear distortion effects can be worse than not doing it, i.e.,

$$E\left[|D_k^{Res}|^2\right] > E\left[|D_k^{Tx}|^2\right]. \tag{5.100}$$

Therefore, the compensation should only take place when the residual nonlinear self-interference is smaller than the nonlinear self-interference in the transmitted signals and when the reliability of the "overall chip" estimates

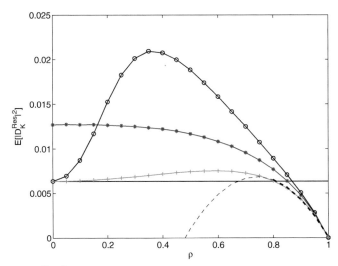

Figure 5.19  $E[|D_k^{Res}|^2]$ as a function of $\rho$ for a normalized clipping level of $s_M/\sigma = 0.5$: given by (5.97) (–o–); given by (5.98) (–*–); given by (5.99) (–+–); true $E[|D_k|^2]$ (—) and given by approximation (5.101), with $x_1 = -15.76$, $x_2 = 23.33$ and $x_3 = -7.55$ (- - -).

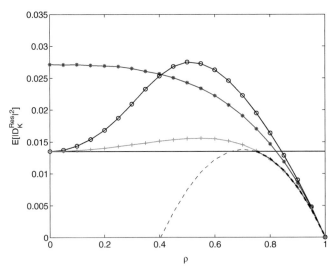

Figure 5.20  As in Fig. 5.19 but with $s_M/\sigma = 1.0$ and $x_1 = -11.36$, $x_2 = 15.92$ and $x_3 = -4.56$.

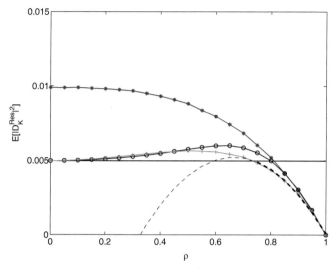

Figure 5.21  As in Fig. 5.19 but with $s_M/\sigma = 2.0$ and $x_1 = -9.48$, $x_2 = 12.70$ and $x_3 = -3.21$.

is above a given threshold. From Figs. 5.19 to 5.21 it is also clear that the best estimates are the ones based on (5.99), which will be considered in the remaining of this chapter. We will assume that

$$E\left[\left|D_k^{Res}\right|^2\right] \approx f(\rho)E\left[\left|D_k^{Tx}\right|^2\right], \tag{5.101}$$

with

$$f(\rho) = x_1\rho^2 + x_2\rho + x_3, \tag{5.102}$$

where $x_1$, $x_2$ and $x_3$ are coefficients that depend on the adopted normalized clipping level $s_M/\sigma$. The optimum values, obtained by simulation, are given in Table 5.1. Approximation (5.101), also included in Figs. 5.19 to 5.21, is used to compute $\eta_k$ and as a threshold to trigger the compensation of non-linear distortion effects (the compensation is performed when $f(\rho) \leq 1$ and $\rho > -x_2/(2x_1)$).

## Implementation Issues

In terms of signal processing complexity, the IB-DFE receiver requires one size-$N$ DFT operation plus a pair of size-$N$ IDFT/DFT operations and a pair of despreading/spreading operations at each iteration to detect a given resolution (except for the first iteration where only one size-$N$ IDFT operation

Table 5.1 Optimum values of $x_1$, $x_2$ and $x_3$.

| $s_M/\sigma$ | $x_1$ | $x_2$ | $x_3$ |
|---|---|---|---|
| 0.5 | −15.76 | 23.33 | −7.55 |
| 1.0 | −11.36 | 15.92 | −4.56 |
| 1.5 | −9.13 | 12.13 | −3.00 |
| 2.0 | −9.48 | 12.70 | −3.21 |

and a despreading operation are required). If estimation and compensation of nonlinear effects are carried out, $I$ pairs of size-$N'$ IDFT/DFT operations for the detection of a given resolution at each iteration are also needed. In the case of the Turbo DFE receiver, the SISO channel decoding needs to be implemented in the detection process, with the soft-input Viterbi algorithm instead of a conventional Viterbi algorithm.

It should be pointed out that the iterative receiver can be simplified with only negligible performance degradation. This can be accomplished by noting that nonlinear distortion effects are mainly due to the spreading codes with higher assigned power. Therefore, in our multi-resolution scenario the lower resolutions (i.e., the ones with higher assigned power) are the ones that should be estimated with higher accuracy so as to obtain accurate estimates of nonlinear distortion effects. This means that if the receiver is only able to detect up to a given resolution (e.g., due to its position relatively to the transmitter and/or due to hardware restrictions), the resolutions which employ sets of spreading codes with lower assigned power could be ignored. Moreover, the estimates for resolutions above the highest we are detecting will be very poor. Naturally, to detect the "higher" resolution (i.e., the resolution with the smaller assigned power) the receiver must detect all resolutions.

Another aspect already mentioned to simplify the receiver is to note that if $\rho_r \approx 1$ for the $r$th resolution at a given iteration, then we already have reliable decisions for that resolution and it can be excluded from the detection process in the next iterations (in fact, we probably already have reliable decisions for all resolutions bellow $r$).

## 5.2.5 Performance Results

In this section we present a set of performance results concerning the improved receivers for multi-resolution broadcasting in DS-CDMA systems employing CP-assisted block transmission techniques, combined with FDE schemes, where an iterative cancelation of deliberate nonlinear distortion effects is carried out. The transmitter (i.e., the BS) simultaneously transmits

Table 5.2 Values of PMEPR and SIR for the transmitted signals.

| $s_M/\sigma$ | PMEPR (dB) C&F Iterations | | | | SIR (dB) C&F Iterations | | | |
|---|---|---|---|---|---|---|---|---|
| | 1 | 2 | 4 | 8 | 1 | 2 | 4 | 8 |
| 0.5 | 4.1 | 3.0 | 2.0 | 1.7 | 8.8 | 7.9 | 7.5 | 7.3 |
| 1.0 | 4.4 | 3.4 | 2.4 | 2.1 | 11.1 | 10.0 | 9.3 | 9.1 |
| 1.5 | 5.0 | 4.0 | 3.2 | 2.9 | 14.6 | 13.0 | 12.2 | 12.0 |
| 2.0 | 5.7 | 4.9 | 4.2 | 4.0 | 19.4 | 17.4 | 16.3 | 15.9 |

data blocks for $N_R$ resolutions. The coded bits associated to each resolution are interleaved before being mapped into QPSK symbols under a Gray mapping rule (the interleaving is performed over five FFT blocks). We consider an orthogonal spreading with $K_r = K/N_R, r = 1, \ldots, N_R$, spreading codes associated to each resolution and the same spreading factor $K = N = 256$ for all spreading codes (this corresponds to a fully loaded system). To reduce the PMEPR of the transmitted signals while maintaining the spectral occupations of conventional DS-CDMA signals, the BS performs an ideal envelope clipping operation, with normalized clipping level $s_M/\sigma$ and an oversampling factor $M_{Tx} = N'/N = 2$, combined with a frequency-domain filtering operation, which can be jointly repeated several times. This procedure allows the PMEPR values and corresponding average SIR values shown in Table 5.2 (see also Figs. 5.7 and 5.14, respectively). The power amplifier at the transmitter is quasi-linear within the (reduced) range of variations of the input signal envelope. The receiver (i.e., the MT) knows the characteristic of the PMEPR-reducing signal processing technique employed by the transmitter.

As mentioned before, we will denote the receiver that employs soft decisions from the FDE in the feedback loop as IB-DFE and the receiver with soft decisions from the SISO channel decoder outputs in the feedback loop as Turbo FDE.

Let us first consider a case where all spreading codes have the same assigned power ($\xi_r = 1, r = 1$), corresponding to a single resolution transmission with a nonlinear transmitter with a normalized clipping level of $s_M/\sigma = 0.5$ and only one C&F operation. Figures 5.22 and 5.24 show the average BER performance for iterations 1 and 4 (naturally, the first iteration corresponds to a linear receiver) for either IB-DFE and Turbo FDE receivers. For the sake of comparisons, we also include the performance for a linear transmitter. Figure 5.22 concerns an ideal AWGN channel and Figs. 5.23 and 5.24 concern the severely time-dispersive channel described in Section 1.3.

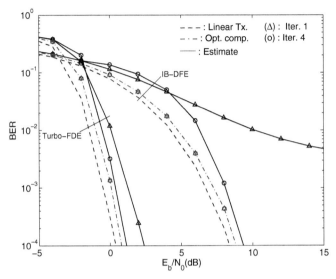

Figure 5.22 Average BER performance for iterations 1 and 4 for an ideal AWGN channel, when $\xi_r = 1, r = 1, \ldots, 4$, for IB-DFE (uncoded BER) and Turbo FDE (coded BER) receivers, when linear and nonlinear transmitters (optimum and estimated nonlinear distortion compensation) with normalized clipping level of $s_M/\sigma = 0.5$ are considered.

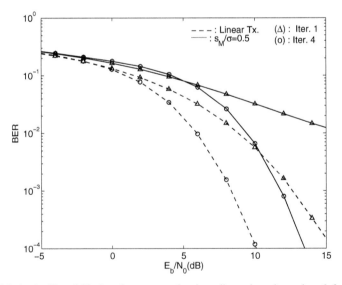

Figure 5.23 As in Fig. 5.22, but for a severely time-dispersive channel and for IB-DFE (uncoded BER) receiver.

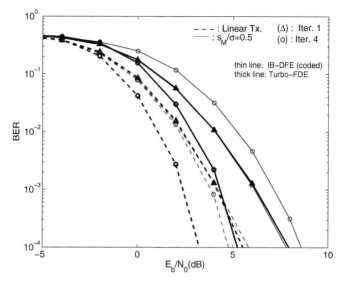

Figure 5.24 As in Fig. 5.23, but for IB-DFE (coded BER) and Turbo FDE (coded BER) receivers.

The duration of the CP is 1/5 of the duration of the useful part of the block (similar results were obtained for other severely time-dispersive channels). From Figs. 5.22 and 5.24 we can observe that the Turbo FDE receiver allows good performance improvements relatively to the linear receiver for both channel types, which can be close to the one obtained with a linear transmitter. However, the uncoded BER performance of the IB-DFE receiver for the time-dispersive channel in Fig. 5.24, remains far from the linear transmitter performance, even after four iterations, and the coded BER performance tends to worsen from iteration to iteration due to error propagation effects resulting from high nonlinear distortion effects.

Let us consider now a multi-resolution scenario with $\xi_1 = 1$, $\xi_2 = 1/2$, $\xi_3 = 1/4$ and $\xi_4 = 1/8$ (i.e., the power assigned to the $r$th resolution is 6 dB below the power assigned to the $(r - 1)$th resolution), assuming again a nonlinear transmitter with $s_M/\sigma = 0.5$ and only one C&F operation. Figures 5.25 to 5.27 concern the same severely time-dispersive channel mentioned above and show, respectively, the uncoded and coded BER performance for the IB-DFE receiver, and the coded BER performance for the Turbo FDE receiver, for iterations 1 and 4. We also include the performance for a linear transmitter. Clearly, the performance degradation due to the nonlinear distortion effects can be very high, especially for low-power

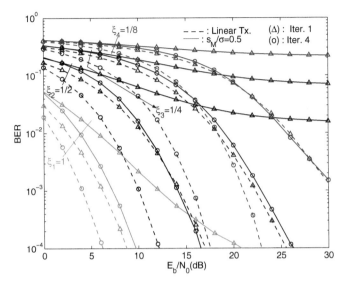

Figure 5.25 Uncoded BER performance for iterations 1 and 4, when $\xi_1 = 1$, $\xi_2 = 1/2$, $\xi_3 = 1/4$ and $\xi_4 = 1/8$, for IB-DFE receiver, when linear and nonlinear transmitters with normalized clipping level of $s_M/\sigma = 0.5$ are considered.

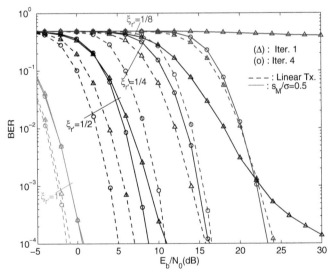

Figure 5.26 Coded BER performance for the IB-DFE receiver, for the same scenario of Fig. 5.25.

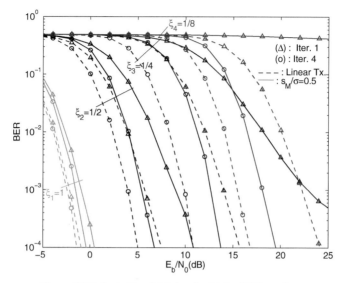

Figure 5.27  As in Fig. 5.26, but for Turbo FDE receiver.

resolutions. As expected, we can also conclude that the Turbo FDE receiver significantly outperforms the IB-DFE.

Let us consider a case where we want a transmission with a very low-PMEPR of the DS-CDMA signals, not only by assuming a very low clipping level, but also by repeating several times the C&F operations to further reduce the PMEPR of the transmitted signals while maintaining the spectral occupation of conventional DS-CDMA schemes. Figure 5.28 shows the average coded BER performance for iterations 1 and 4 for Turbo FDE receiver with 1, 2, 4 or 8 C&F iterations at the transmitter and a normalized clipping level of $s_M/\sigma = 0.5$. From this figure it is clear that the performance degradation associated to several C&F operations is relatively small when Turbo FDE receivers with estimation and cancelation of nonlinear distortion effects are employed, in spite of the considerable PMEPR reduction of the transmitted signals.

## 5.3  Nonlinear Effects in MC-CDMA Signals

This section presents an extension of the transmitter and receiver structures for the downlink transmission of low-PMEPR DS-CDMA signals presented in the previous section, to the uplink transmission of low-PMEPR MC-

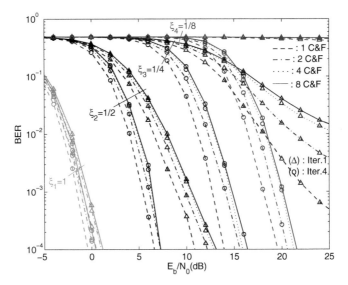

Figure 5.28 Coded BER performance for iterations 1 and 4, when $\xi_1 = 1$, $\xi_2 = 1/2$, $\xi_3 = 1/4$ and $\xi_4 = 1/8$, for Turbo FDE receiver, with 1, 2, 4 or 8 C&F iterations at the transmitter and a normalized clipping level of $s_M/\sigma = 0.5$.

CDMA signals, employing clipping techniques combined with frequency-domain filtering. Although the approach considered in the following concerns the uplink transmission of MC-CDMA system, it should be pointed out that a similar approach can also be applied for the downlink transmission of MC-CDMA signals, since the transmitted signals also results from the sum of several components associated to different spreading codes and, consequently, leads to high PMEPR signals.

### 5.3.1 Transmitter Structure

Let us consider the uplink transmission in MC-CDMA systems employing frequency-domain spreading. The frequency-domain block of chips to be transmitted by the $p$th MT is $\{S_{k,p}; k = 0, 1, \ldots, N - 1\}$ with $S_{k,p}$ given by (2.63).

To reduce the envelope fluctuations of the transmitted signals, we employ the transmitter structure depicted in Fig. 5.29. As for the DS-CDMA scheme, the nonlinear device, characterized by $g_C(\cdot)$ and given by (5.12), is modeled as a bandpass memoryless nonlinearity [84] operating on an oversampled

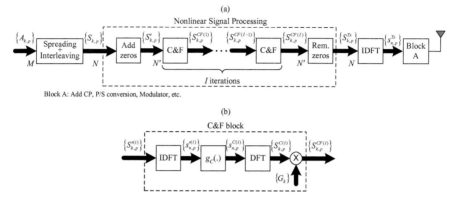

Figure 5.29  (a) MC-CDMA nonlinear transmitter structure and (b) detail of the C&F block.

version of the MC-CDMA block, i.e.,

$$s_{n,p}^{C} = g_C(|s_{n,p}'|) \exp(j \arg(s_{n,p}')). \qquad (5.103)$$

As was already explained, to mitigate the PMEPR regrowth after the filtering operation, the C&F operations can be repeated, in an iterative way, $I$ times with

$$S_{k,p}'^{(i)} = \begin{cases} S_{k,p}', & i = 1 \\ S_{k,p}^{CF(i-1)}, & i > 1. \end{cases} \qquad (5.104)$$

The block of modified samples to be transmitted $\{s_{n,p}^{Tx}; n = 0, 1, \ldots, N - 1\}$ is obtained after computing the corresponding IDFT.

The statistic characterization of the transmitted signals for the DS-CDMA system, described in Section 5.2.2, taking advantage of the Gaussian nature of the signal at the input of the memoryless nonlinear device, can be extended for each of the MC-CDMA transmitters. In fact, when the number of subcarriers is high ($N \gg 1$) the time-domain coefficients $s_{n,p}'$ can be approximately regarded as samples of a complex Gaussian process. Having in mind (5.42) and the signal processing chain in Fig. 5.29(b), the frequency-domain block $\{S_{k,p}^{Tx} = S_{k,p}^{C} G_k; k = 0, 1, \ldots, N' - 1\}$ can obviously be decomposed into useful and nonlinear self-interference components:

$$S_{k,p}^{Tx} = \alpha_{k,p} S_{k,p}' G_k + D_{k,p} G_k, \qquad (5.105)$$

where $\{D_{k,p}; k = 0, 1, \ldots, N' - 1\}$ is the DFT of $\{d_{n,p}; n = 0, 1, \ldots, N' - 1\}$. Clearly, $E[D_{k,p}] = 0$ and

$$E\left[D_{k,p} D_{k',p}^*\right] = \begin{cases} N' G_d(k), & k = k' \\ 0, & \text{otherwise}, \end{cases} \qquad (5.106)$$

$k, k' = 0, 1, \ldots, N' - 1$, where $\{G_d(k); k = 0, 1, \ldots, N' - 1\}$ denotes the DFT of the block $\{R_d(n); n = 0, 1, \ldots, N' - 1\}$. Moreover, $D_{k,p}$ exhibits quasi-Gaussian characteristics for any $k$, provided that the number of subcarriers is high enough. Clearly,

$$E\left[S^C_{k,p} S^{C*}_{k',p}\right] = \begin{cases} N' G^C_s(k), & k = k' \\ 0, & \text{otherwise,} \end{cases} \tag{5.107}$$

where $\{G^C_s(k) = |\alpha_{k,p}|^2 G_s(k) + G_d(k); k = 0, 1, \ldots, N' - 1\}$ denotes the DFT of $\{R^C_s(n); k = 0, 1, \ldots, N' - 1\}$, given by (see Appendix F)

$$E\left[s^C_{n,p} s^{C*}_{n',p}\right] = R^C_s(n - n')$$

$$= \sum_{\gamma=0}^{+\infty} 2P_{2\gamma+1} \frac{(R_s(n - n'))^{\gamma+1} (R^*_s(n - n'))^{\gamma}}{(R_s(0))^{2\gamma+1}}, \tag{5.108}$$

and $\{G_s(k); k = 0, 1, \ldots, N' - 1\}$ the DFT of $\{R_s(n); n = 0, 1, \ldots, N' - 1\}$. Therefore,

$$E\left[S^{Tx}_{k,p} S^{Tx*}_{k',p}\right] = \begin{cases} |G_k|^2 E\left[\left|S^C_{k,p}\right|^2\right] = N' |G_k|^2 G^C_s(k), & k = k' \\ 0, & \text{otherwise.} \end{cases} \tag{5.109}$$

Clearly, the power of the useful and nonlinear self-interference components of the transmitted signals are

$$P^S_{Tx} = \sum_{k=0}^{N'-1} E\left[\left|\alpha_{k,p} S'_{k,p} G_k\right|^2\right] \tag{5.110}$$

and

$$P^I_{Tx} = \sum_{k=0}^{N'-1} E\left[\left|D_{k,p} G_k\right|^2\right], \tag{5.111}$$

respectively. We can also define the power of the nonlinear self-interference component in the in-band region as

$$P^I_{Tx,IB} = \sum_{k \text{ in-band}} E\left[\left|D_{k,p} G_k\right|^2\right]. \tag{5.112}$$

When $G_k = 1$ for the $N$ in-band subcarriers, $P^S_{Tx} = P^S_{NL}$. If we also have $G_k = 0$ for the $N' - N$ out-of-band subcarriers then $P^I_{Tx,IB} = P^I_{Tx}$.

The SIR for the transmitted signals is

$$\mathrm{SIR}_{Tx} = \frac{P_{Tx}^S}{P_{Tx}^I} \leq \mathrm{SIR}_{NL} = \frac{P_{NL}^S}{P_{NL}^I}, \tag{5.113}$$

where $\mathrm{SIR}_{NL}$ denotes the SIR at the output of the nonlinear device; the SIR for the in-band region is

$$\mathrm{SIR}_{Tx,IB} = \frac{P_{Tx}^S}{P_{Tx,IB}^I}. \tag{5.114}$$

We can also define a SIR for each subcarrier, given by

$$\mathrm{SIR}_k = \frac{E\left[\left|\alpha_{k,p} S'_{k,p}\right|^2\right]}{E\left[\left|D_{k,p}\right|^2\right]}. \tag{5.115}$$

Without oversampling, (5.29) leads to $R_s(n - n') = 2\sigma^2 \delta_{n,n'}$ and, from (5.108),

$$R_s^C(n - n') = \begin{cases} 2 \displaystyle\sum_{\gamma=0}^{+\infty} P_{2\gamma+1}, & n = n' \\ 0, & \text{otherwise.} \end{cases} \tag{5.116}$$

Therefore,

$$\mathrm{SIR}_k = \frac{P_1}{\displaystyle\sum_{\gamma=1}^{+\infty} P_{2\gamma+1}} = \mathrm{SIR}_{NL} = \mathrm{SIR}_{Tx} = \mathrm{SIR}_{Tx,IB}, \tag{5.117}$$

which is independent of $k$, when $M_{Tx} = 1$. For $M_{Tx} > 1$ (i.e., when $N' > N$), $R_s(n - n') \neq 2\sigma^2 \delta_{n,n'}$ and $\mathrm{SIR}_k$ is a function of $k$, since

$$E\left[\left|D_{k,p}\right|^2\right] = 2\sigma_{D,p}^2(k)\delta_{k,k'} \tag{5.118}$$

depends also on $k$. For a transmitter with successive C&F operations, $\sigma_{D,p}^2(k)$ has to be obtained by simulation.

## 5.3.2 Receiver Design

### Receiver Structure
It was shown in [93, 94] that we can improve significantly the performance of OFDM schemes introduced into nonlinear devices by employing a receiver with iterative cancelation of nonlinear distortion effects. This concept,

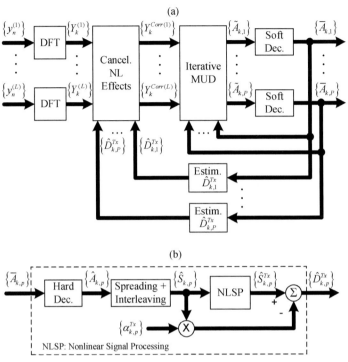

Figure 5.30 (a) Iterative receiver with estimation and compensation of nonlinear distortion effects and (b) detail of the nonlinear distortion estimation blocks.

already applied on the multicode DS-CDMA system in the previous section, can be extended to MC-CDMA, leading to the receiver structure of Fig. 5.30(a), where it is assumed that the BS has several receive antennas, so as to reduce the transmit power requirements of each MT. Once again, the basic idea behind this receiver is to use the estimates of the nonlinear distortion $D_{k,p}^{Tx}$, $\hat{D}_{k,p}^{Tx}$, provided by the preceding iteration to remove the nonlinear distortion effects in the received samples.

We will define the length-$KLP$ vector $\mathbf{D}^{Tx}(k)$ as the concatenation of

$$
\begin{bmatrix}
\mathbf{D}_k^{Tx} \\
\mathbf{D}_{k+M}^{Tx} \\
\vdots \\
\mathbf{D}_{k+(K-1)M}^{Tx}
\end{bmatrix}
\tag{5.119}
$$

$L$ times, where

$$\mathbf{D}_k^{Tx} = \begin{bmatrix} D_{k,1}^{Tx} \\ \vdots \\ D_{k,P}^{Tx} \end{bmatrix}, \tag{5.120}$$

and $\hat{\mathbf{D}}^{Tx}(k)$ its estimate (obtained from the previous iteration). We also define the $KLP \times KL$ matrix

$$\mathbf{H}^{Ch}(k) = \begin{bmatrix} \begin{bmatrix} H_k^{Ch(1)} & \cdots & \mathbf{0}_{1\times P} \\ \mathbf{0}_{1\times P} & \cdots & \mathbf{0}_{1\times P} \\ \vdots & \ddots & \vdots \\ \mathbf{0}_{1\times P} & \cdots & H_{k+(K-1)M}^{Ch(1)} \end{bmatrix} & \cdots & \mathbf{0}_{K\times KP} \\ \vdots & \ddots & \vdots \\ \mathbf{0}_{K\times KP} & \cdots & \begin{bmatrix} H_k^{Ch(L)} & \cdots & \mathbf{0}_{1\times P} \\ \mathbf{0}_{1\times P} & \cdots & \mathbf{0}_{1\times P} \\ \vdots & \ddots & \vdots \\ \mathbf{0}_{1\times P} & \cdots & H_{k+(K-1)M}^{Ch(L)} \end{bmatrix} \end{bmatrix}^{T} \tag{5.121}$$

with

$$\mathbf{H}_k^{Ch(l)} = \begin{bmatrix} H_{k,1}^{Ch(l)} & H_{k,2}^{Ch(l)} & \cdots & H_{k,P}^{Ch(l)} \end{bmatrix}. \tag{5.122}$$

Therefore, the received frequency-domain block vector, $\mathbf{Y}(k)$, which for nonlinear transmitters is

$$\mathbf{Y}(k) = \mathbf{H}^T(k)\mathbf{U}(k)\mathbf{A}(k) + \mathbf{H}^{Ch^T}(k)\mathbf{D}^{Tx}(k) + \mathbf{N}(k), \tag{5.123}$$

is replaced by the corrected block vector $\mathbf{Y}^{Corr}(k)$, given by

$$\begin{aligned} \mathbf{Y}^{Corr}(k) &= \mathbf{Y}(k) - \mathbf{H}^{Ch^T}(k)\hat{\mathbf{D}}^{Tx}(k) \\ &= \mathbf{H}^T(k)\mathbf{U}(k)\mathbf{A}(k) + \mathbf{H}^{Ch^T}(k)\mathbf{D}^{Res}(k) + \mathbf{N}(k) \\ &= \mathbf{H}^{Use^T}(k)\mathbf{A}(k) + \mathbf{H}^{Ch^T}(k)\mathbf{D}^{Res}(k) + \mathbf{N}(k) \end{aligned} \tag{5.124}$$

where

$$\mathbf{U}(k) = \operatorname{diag}(\alpha_{k,1}^{Tx}, \ldots, \alpha_{k,P}^{Tx}), \tag{5.125}$$

$$\mathbf{H}^{Use}(k) = \mathbf{U}(k)\mathbf{H}(k) \tag{5.126}$$

and

$$\mathbf{D}^{Res}(k) = \mathbf{D}^{Tx}(k) - \hat{\mathbf{D}}^{Tx}(k) \tag{5.127}$$

is the residual nonlinear self-distortion vector.

**Derivation of the Receiver Parameters**

To obtain the feedforward and feedback matrices, $\mathbf{F}(k)$ and $\mathbf{B}(k)$, respectively, we can use the same approach as we did for linear transmitters in Section 4.3.2. These matrices are chosen so as to maximize the $\{\text{SINR}_p; p = 1, \ldots, P\}$ given by (4.114), for all MTs, at a particular iteration. This maximization problem is equivalent to the minimization of the MSE, $E\left[|\mathbf{\Theta}(k)|^2\right]$, with

$$\mathbf{\Theta}(k) = \tilde{\mathbf{A}}(k) - \mathbf{A}(k), \tag{5.128}$$

conditioned to $\gamma_p \alpha_{k,p}^{Tx} = 1$, $p = 1, \ldots, P$, which can be solved by employing the Lagrange multipliers' method.

The samples vector $\tilde{\mathbf{A}}(k)$ is given by

$$\tilde{\mathbf{A}}(k) = \mathbf{F}^T(k)\mathbf{Y}^{Corr}(k) - \mathbf{B}^T(k)\overline{\mathbf{A}}(k). \tag{5.129}$$

Substituting (4.111) and (5.124) in (5.129) we obtain

$$\tilde{\mathbf{A}}(k) = \left(\mathbf{F}^T(k)\mathbf{H}^{Use^T}(k) - \mathbf{B}^T(k)\mathbf{P}^2\right)\mathbf{A}(k) - \mathbf{B}^T(k)\mathbf{P}\mathbf{\Delta}(k)$$
$$+ \mathbf{F}^T(k)\left(\mathbf{H}^{Ch^T}(k)\mathbf{D}^{Res}(k) + \mathbf{N}(k)\right), \tag{5.130}$$

resulting an overall error vector given by

$$\mathbf{\Theta}(k) = \left(\mathbf{F}^T(k)\mathbf{H}^{Use^T}(k) - \mathbf{B}^T(k)\mathbf{P}^2 - \mathbf{I}_P\right)\mathbf{A}(k) - \mathbf{B}^T(k)\mathbf{P}\mathbf{\Delta}(k)$$
$$+ \mathbf{F}^T(k)\left(\mathbf{H}^{Ch^T}(k)\mathbf{D}^{Res}(k) + \mathbf{N}(k)\right). \tag{5.131}$$

After some mathematical manipulation the MSE is given by

$$E[|\mathbf{\Theta}(k)|^2] = 2\sigma_A^2\left(\mathbf{F}^H(k)\mathbf{H}^{Use^H}(k)\mathbf{H}^{Use}(k)\mathbf{F}(k) + \mathbf{B}^H(k)\mathbf{P}^4\mathbf{B}(k) + \mathbf{I}_P\right.$$
$$- \mathbf{F}^H(k)\mathbf{H}^{Use^H}(k)\mathbf{P}^2\mathbf{B}(k) - \mathbf{F}^H(k)\mathbf{H}^{Use^H}(k)$$
$$- \mathbf{B}^H(k)\mathbf{P}^2\mathbf{H}^{Use}(k)\mathbf{F}(k) + \mathbf{B}^H(k)\mathbf{P}^2 - \mathbf{H}^{Use}(k)\mathbf{F}(k)$$
$$+ \mathbf{P}^2\mathbf{B}(k) + \mathbf{B}^H(k)\mathbf{P}^2(\mathbf{I}_P - \mathbf{P}^2)\mathbf{B}(k))$$
$$+ \mathbf{F}^H(k)\mathbf{H}^{Ch^H}(k)\mathbf{R}_D(k)\mathbf{H}^{Ch}(k)\mathbf{F}(k) + 2\sigma_N^2\mathbf{F}^H(k)\mathbf{F}(k)$$
$$\tag{5.132}$$

where

$$\mathbf{R}_D(k) = E\left[\mathbf{D}^{Res^*}(k)\mathbf{D}^{Res^T}(k)\right]. \tag{5.133}$$

Defining the matrix of Lagrange functions

$$\mathbf{J} = E\left[|\mathbf{\Theta}(k)|^2\right] + \left(\mathbf{\Gamma}(k)\mathbf{U}(k) - \mathbf{I}_P\right)\mathbf{\Lambda}, \tag{5.134}$$

and assuming that the optimization is carried out under $\mathbf{\Gamma}(k)\mathbf{U}(k) = \mathbf{I}_P$, the optimum feedforward and feedback matrices can be obtained by solving the following set of equations:

- $\nabla_{\mathbf{F}(k)}\mathbf{J} = 0 \Leftrightarrow 4\sigma_A^2\left(\mathbf{H}^{Use^H}(k)\mathbf{H}^{Use}(k)\mathbf{F}(k) - \mathbf{H}^{Use^H}(k)\mathbf{P}^2\mathbf{B}(k)\right.$

$$\left. - \mathbf{H}^{Use^H}(k)\right) + 2\mathbf{H}^{Ch^H}(k)\mathbf{R}_D(k)\mathbf{H}^{Ch}(k)\mathbf{F}(k)$$

$$+4\sigma_N^2\mathbf{F}(k) + \frac{2}{M}\mathbf{H}^{Use^H}(k)\mathbf{\Lambda} = 0$$

$$\Leftrightarrow \mathbf{H}^{Use^H}(k)\mathbf{H}^{Use}(k)\mathbf{F}(k) - \mathbf{H}^{Use^H}(k)\mathbf{P}^2\mathbf{B}(k) - \mathbf{H}^{Use^H}(k)$$

$$+\frac{1}{2\sigma_A^2}\mathbf{H}^{Ch^H}(k)\mathbf{R}_D(k)\mathbf{H}^{Ch}(k)\mathbf{F}(k) + \beta\mathbf{F}(k)$$

$$+\frac{1}{2M\sigma_A^2}\mathbf{H}^{Use^H}(k)\mathbf{\Lambda} = 0, \tag{5.135}$$

- $\nabla_{\mathbf{B}(k)}\mathbf{J} = 0 \Leftrightarrow \mathbf{P}^4\mathbf{B}(k) - \mathbf{P}^2\mathbf{H}^{Use}(k)\mathbf{F}(k) + \mathbf{P}^2 + \mathbf{P}^2(\mathbf{I}_P - \mathbf{P}^2)\mathbf{B}(k) = 0$

$$\Leftrightarrow \mathbf{B}(k) = \mathbf{H}^{Use}(k)\mathbf{F}(k) - \mathbf{I}_P \tag{5.136}$$

(5.136) is the optimum feedback matrix;

- $\nabla_{\mathbf{\Lambda}}\mathbf{J} = 0 \Leftrightarrow \mathbf{\Gamma}(k)\mathbf{U}(k) = \mathbf{I}_P \tag{5.137}$

(5.137) is the conditions under which the optimization is carried out.

By replacing (5.136) in (5.135), the optimum feedforward matrix $\mathbf{F}(k)$ can be written as

$$\mathbf{F}(k) = \left[\mathbf{H}^{Use^H}(k)\left(\mathbf{I}_P - \mathbf{P}^2\right)\mathbf{H}^{Use}(k) + \frac{1}{2\sigma_A^2}\mathbf{H}^{Ch^H}(k)\mathbf{R}_D(k)\mathbf{H}^{Ch}(k)\right.$$

$$\left. + \beta\mathbf{I}_{KL}\right]^{-1}\mathbf{H}^{Use^H}(k)\mathbf{Q}, \tag{5.138}$$

where the constant normalization matrix

$$\mathbf{Q} = \text{diag}(Q_1, \ldots, Q_P) = \mathbf{I}_P - \mathbf{P}^2 - \frac{1}{2M\sigma_A^2}\mathbf{\Lambda} \tag{5.139}$$

ensures that $\mathbf{\Gamma}(k)\mathbf{U}(k) = \mathbf{I}_P$.

## Derivation of the Nonlinear Distortion Estimates

For the first iteration, since $\hat{\mathbf{D}}^{Tx}(k) = 0$ and $\mathbf{D}^{Res}(k) = \mathbf{D}^{Tx}(k)$, the non-zero elements of $\mathbf{R}_D(k)$ are of the type

$$E\left[\left|D_{k,p}^{Tx}\right|^2\right] = 2\sigma_{D,p}^2(k). \tag{5.140}$$

For the remaining iterations, $\mathbf{D}^{Res}(k) = \mathbf{D}^{Tx}(k) - \hat{\mathbf{D}}^{Tx}(k)$, where the elements $\{\hat{D}_{k,p}^{Tx}; k = 0, 1, \ldots, N - 1\}$ of $\hat{\mathbf{D}}^{Tx}(k)$, obtained from the previous iteration, can be estimated, as in Section 5.2.4, from $\{\hat{A}_{k,p}; k = 0, 1, \ldots, M - 1\}$ as follows (see Fig. 5.30(b)): $\{\hat{A}_{k,p}; k = 0, 1, \ldots, M - 1\}$ is re-spread to generate an estimate of the "block to be transmitted" $\{\hat{S}_{k,p}; k = 0, 1, \ldots, N - 1\}$; $\{\hat{S}_{k,p}; k = 0, 1, \ldots, N - 1\}$ is submitted to a replica of the nonlinear signal processing scheme employed in the $p$th transmitter so as to form the "transmitted block estimate" $\{\hat{S}_{k,p}^{Tx}; k = 0, 1, \ldots, N - 1\}$; finally, $\hat{D}_{k,p}^{Tx}$ is given by

$$\hat{D}_{k,p}^{Tx}\bigg|_{\{\hat{A}_{k,p}\}} = \hat{S}_{k,p}^{Tx}\bigg|_{\{\hat{A}_{k,p}\}} - \alpha_{k,p}^{Tx}\hat{S}_{k,p}. \tag{5.141}$$

For the same reasons explained in Section 5.2.4 (i.e., to avoid propagation errors), we will consider the weighted estimates, expressed by (5.99), which is this case are given by

$$\hat{D}_{k,p}^{Tx} = \rho_p \, \hat{D}_{k,p}^{Tx}\bigg|_{\{\hat{A}_{k,p}\}}. \tag{5.142}$$

Similarly to (5.101), we will also consider the approximation

$$E\left[\left|D_{k,p}^{Res}\right|^2\right] \approx f(\rho)E\left[\left|D_{k,p}^{Tx}\right|^2\right], \tag{5.143}$$

with $f(\rho)$ given by (5.102), to compute $\mathbf{R}_D(k)$ and as a threshold to trigger the compensation of the nonlinear distortion effects.

## Implementation Issues

Relatively to the MUD receivers for linear transmitters discussed in Section 4.3, the MUD receivers for nonlinear transmitters have the additional complexity inherent to the estimation and compensation of nonlinear effects. Therefore, besides the set of DFT/IDFT and despreading/spreading operations, as well as the computation charge inherent to the feedforward and feedback coefficients for the receivers for linear transmitters, $IP$ pairs of size-$N'$ DFT/IDFT operations plus $P$ spreading operations for the detection of each MT, at each iteration, are also needed to estimate the nonlinear distortion when nonlinear transmitters are used.

When compared with a conventional MRC-PIC receiver [101], our receivers are more complex, with an additional implementation complexity coming essentially from the computation of the feedforward coefficients and from the extra size-$N$ DFT/IDFT operations required when estimation and compensation of nonlinear effects are employed. However, the main computational effort can be the one inherent to SISO channel decoding, which is something that is also required in MRC-PIC receivers with turbo decoding. Moreover, MRC-PIC receivers have poor performance for high system load, while the Turbo MUD receiver proposed herein has good performance even for fully loaded systems, as we will see in the following.

### 5.3.3 Performance Results

In this section we present a set of performance results concerning the iterative receiver structures proposed for the uplink of MC-CDMA systems with nonlinear transmitters. We consider $M = 32$ data symbols for each user, corresponding to blocks with length $N = KM = 256$, plus an appropriate CP. The receiver (i.e., the BS) knows the characteristics of the PMEPR-reducing signal processing technique employed by each MT and employs $L$ uncorrelated receive antennas, for diversity purposes. Unless otherwise stated, we consider $P = K = 8$ MTs, corresponding to a fully loaded scenario and $\xi_p = 1$ for all MTs, i.e., we have perfect "average power control".

As before, we will denote the receiver with soft decisions from the multiuser detector employed in the feedback loop as IMUD receiver and the receiver with soft decisions from the channel decoder outputs as Turbo MUD receiver.

Let us first consider an uncoded case where we have nonlinear transmitters at each MT with a normalized clipping level, identical for all MTs, of $s_M/\sigma = 1.0$ and $s_M/\sigma = 0.5$. Figures 5.31 and 5.32 show the average uncoded BER performance (i.e., the average over all MTs) for iterations 1 and 4 when $L = 1$ or 2, respectively. For the sake of comparisons, we also include the performance for a linear transmitter and the SU performance defined in (2.57). From Figs. 5.31 and 5.32 it is clear that the iterative receiver allows significant performance improvements relatively to the linear receiver, although, for $L = 1$, the performance are far from the SU performance, even after four iterations. Moreover, our simulation results showed that we can approach the SU performance, when we have diversity.

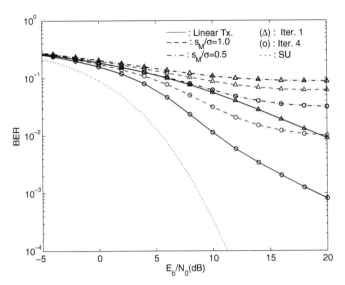

Figure 5.31 Average uncoded BER performance for iterations 1 and 4 with $L = 1$, when linear and nonlinear transmitters with normalized clipping level of $s_M/\sigma = 1.0$ and 0.5 are considered.

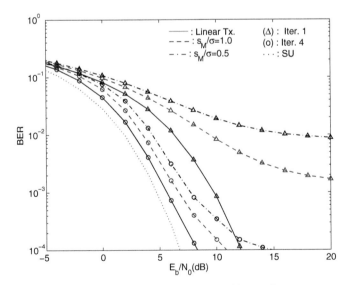

Figure 5.32 As in Fig. 5.31 but with $L = 2$.

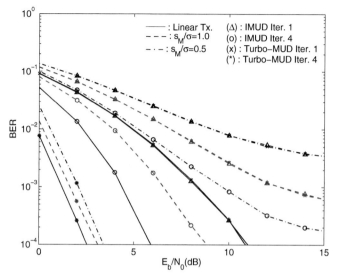

Figure 5.33 Average coded BER performance for iterations 1 and 4 for either IMUD and Turbo MUD receivers with $L = 1$, when linear and nonlinear transmitters with normalized clipping level of $s_M/\sigma = 1.0$ and 0.5 are considered.

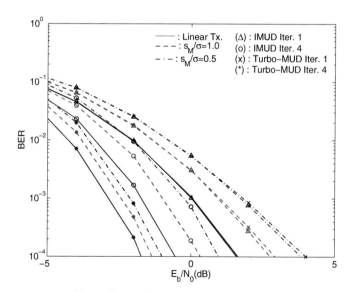

Figure 5.34 As in Fig. 5.33 but with $L = 2$.

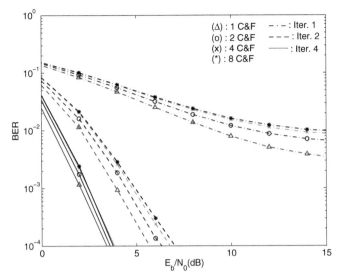

Figure 5.35 Average coded BER performance for iterations 1, 2 and 4 for Turbo MUD receiver with $L = 1$, with 1, 2, 4 or 8 C&F iterations at the transmitter and a normalized clipping level of $s_M/\sigma = 0.5$.

Let us consider now the impact of channel coding by assuming again that we have nonlinear transmitters at each MT with normalized clipping levels of $s_M/\sigma = 1.0$ and $s_M/\sigma = 0.5$. Figures 5.33 and 5.34 show the average coded BER performance for iterations 1 and 4, again for $L = 1$ or 2, respectively, for either IMUD and Turbo MUD receivers. As expected, the channel coding leads to significant performance improvements. Moreover, it is clear that the performance of the linear receiver is very poor, with high irreducible error floors due to the nonlinear distortion effects. This is especially serious for the case where $L = 1$. As we increase the number of iterations and/or we increase $L$ improves significantly the performance, which can be close to the one obtained with linear transmitters if $L > 1$. We can also observe that the Turbo MUD outperforms the IMUD, especially when $L = 1$.

Let us consider a case where we want a transmission with a very low-PMEPR of the MC-CDMA signals, not only by assuming a very low clipping level at each MT, but also by repeating several times the C&F operations at each MT to further reduce the PMEPR of the transmitted signals while maintaining the spectral occupation of conventional MC-CDMA schemes (see Table 5.2). Figure 5.35 shows the average coded BER performance for iterations 1, 2 and 4 for Turbo MUD receiver with $L = 1$, with 1, 2, 4

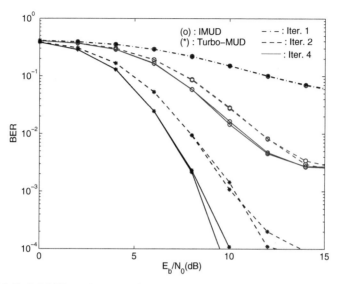

Figure 5.36 Coded BER performance for each $C_L$ user as a function of the $E_b/N_0$ of $C_H$ users, when $P = K = 4$, $L = 1$ and $s_M/\sigma = 1.0$, for either IMUD or Turbo MUD receivers.

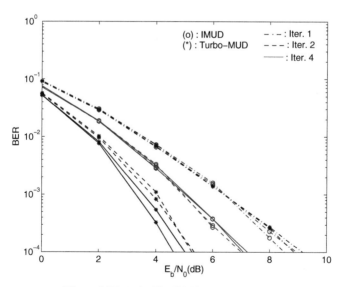

Figure 5.37 As in Fig. 5.36 but for $C_H$ users.

or 8 C&F iterations at the transmitter and a normalized clipping level of $s_M/\sigma = 0.5$. From this figure it is clear that the performance degradation associated to several C&F operations is very small when Turbo MUD receivers with estimation and cancelation of nonlinear distortion effects are employed.

Finally, let us consider now a fully loaded scenario with $K = P = 4$, $L = 1$ and a normalized clipping level of $s_M/\sigma = 1.0$ where the signals associated to different users have different average power at the receiver. We will consider two classes of users, denoted by $C_L$ and $C_H$, with two users in each class, where the average power of $C_H$ users is 6dB above the average power of $C_L$ users. Clearly, the $C_L$ users face strong interference conditions. The coded BER performance for each $C_L$ and $C_H$ user as a function of $E_b/N_0$ of $C_H$ users is shown in Figs. 5.36 and 5.37, respectively, for either IMUD and Turbo MUD receivers. Once again, the iterative receiver allows significant performance gains, with the Turbo MUD receiver outperforming the IMUD, especially for $C_L$ users.

# 6

---

# Conclusions and Future Work

---

## 6.1 Conclusions

The main objective of this work was the design of sophisticated frequency-domain receiver structures combining MUD techniques with iterative signal detection/decoding techniques based on the IB-DFE concept in the context of a CP-assisted blockwise transmission for DS-CDMA and MC-CDMA systems. Joint turbo equalization and MUD receiver structures, eventually combined with estimation and cancelation of nonlinear distortion effects, with different complexity/performance tradeoffs were proposed, suitable to scenarios with high interference levels and strongly time-dispersive channels. Both the downlink and uplink transmissions were considered.

Chapter 2 introduced the basic principles of MC and SC modulations. It was shown how SC modulations could benefit from FDE techniques by taking advantage of the CP-assisted block transmission approach and enabling an efficient use of FFT-based implementations, as with current MC-based OFDM modulations. These concepts were then extended to the CDMA schemes MC-CDMA and DS-CDMA. As with OFDM and SC-FDE, the use of linear frequency-domain receiver implementations are especially interesting for high data rate CP-assisted block transmissions over severely time-dispersive channels (due to FFT-based implementations), with much lower complexity then the optimum receivers. However, when an MMSE-FDE is used instead of a ZF-FDE to avoid noise enhancement effects, severe residual interference levels can be expected for high system load and/or when different users have different assigned powers, leading to poor BER performance.

Although the issues presented in Chapter 2 are not original, they greatly facilitate the understanding and motivation of the original subjects proposed in Chapters 3, 4 and 5. All the original work presented in these chapters was published in several international journals [32, 51, 76–79] and conference proceedings [31, 48–50, 55–59, 70–75, 80].

193

Chapter 3 presented receiver structures for the downlink transmission of DS-CDMA and MC-CDMA systems based on the IB-DFE techniques as an alternative to conventional linear FDE techniques. With these IB-DFE techniques, the result of the first iteration corresponds to those of conventional linear MMSE-FDE technique; the subsequent iterations provide a performance enhancement, thanks to the iterative cancelation of residual interference, that can be close to the MFB performance, especially for DS-CDMA schemes. To improve the performance, an efficient turbo FDE receiver design based on the IB-DFE concept was also defined. In this case, as conventional turbo equalizer, the "soft decisions" from the FDE outputs are replaced by "soft decisions" from the channel decoder outputs in the feedback loop. As expected, with this "turbo FDE" receiver, significant performance improvements relatively to the uncoded case are achieved, especially for the MC-CDMA system, whose performance approaches the MFB, confirming that channel coding when combined with appropriate interleaving can compensate the worse uncoded performance.

Chapter 4 was dedicated to the design of iterative frequency-domain MUD receivers for the uplink transmission of DS-CDMA and MC-CDMA systems. The proposed receivers combine turbo equalization and MAI cancelation techniques. Both SIC and PIC structures were considered as well as an extension to MIMO systems allowing significant increase in the system's spectral efficiency. The performance results showed that the proposed receivers, especially the Turbo MUD, can have performance close to the SU/MFB one, even for fully loaded systems and severely time-dispersive channels. Moreover, the Turbo MUD receiver can also cope with overloaded scenarios, where the number of users exceeds the spreading factor, although with some performance degradation and a higher number of iterations.

Chapter 5 considered the use of nonlinear transmitters for multi-resolution DS-CDMA systems and for the uplink transmission in MC-CDMA systems. The transmitters employ suitable clipping techniques combined with frequen-cy-domain filtering, jointly performed in an iterative way, so as to reduce the envelope fluctuations and PMEPR of the transmitted signals. Improved receiver structures were proposed, able to perform the detection and an iterative estimation and threshold-based cancelation of deliberate nonlinear distortion effects inherent to the transmitted signals. The performance results show that significant performance improvement can be obtained, with performance close to the one of linear receivers, after just a few iterations even for severely time-dispersive channels and/or in the presence of strong nonlinear effects. In fact, we can have transmitted signals with PMEPR values as low as 1.7 dB. Moreover, the results also show that the use of channel decoder outputs in the receiver's feedback loop allows a significant perform-ance improvement at low and moderate SNR, which is especially important for scenarios where different power levels are assigned to different resolu-tions/transmitted signals. Although the proposed receivers require additional implementation complexity, especially for the Turbo-MUD receiver, it was pointed out that the implementation effort is concentrated in the BS, where increased power consumption and cost are not so critical. Having in mind the benefits of using efficient, low-cost power amplification at the MTs, the proposed techniques could mean an important implementation advantage for the MTs.

## 6.2 Future Work

There are several aspects concerning the transmit/receive structures studied in this book that were not addressed and can be the subject of future work. Among others, the following extensions to this work can be the subject of further investigation:

- **Channel frequency response estimation issues**
  All performance results presented in this work were obtained by assum-ing a perfect knowledge of the channel conditions. However, in practice, the channel frequency response is not known and must be estimated at the receiver. Therefore, an important issue to be further investigated is the impact of an imperfect channel estimation on the receivers' BER performance proposed herein, as well as suitable channel estimation techniques.

- **Synchronization issues**
  The proposed receiver structures are assumed to work in perfect time and frequency synchronization scenarios between the BS and the MT. Typic-ally, these synchronization requirements are ensured though a feedback channel, from the BS to each MT. As previously referred, perfect time synchronization between the block associated to different MTs is not re-quired since some time mismatches can be absorbed by the CP. However, some residual time and/or frequency errors are unavoidable, especially if a low-rate feedback channel is intended. Although some investigation was carried out on this subject that was not included in this book, partic-ularly on the receiver design for the uplink transmission in asynchronous

CP-assisted DS-CDMA system [58], further work needs to be developed on this topic, namely, for the MC-CDMA system.

- **Analysis on the convergence of the iterative procedure of the receiver**
  The simulation results presented in this work and many others obtained through out this investigation show that the iterative procedure inherent to the nonlinear equalization techniques employed in the proposed receivers structures is very robust. However, a detailed study on the convergence properties of the iterative procedure, eventually resorting to EXIT (Extrinsic Information Transfer) charts, is another open issue to investigate.

# Appendices

# A

# Channel Characterization

In a multipath propagation environment the complex envelope of the received signal can be regarded as the sum of different replicas of the complex envelope of the transmitted signal with different delays and complex amplitudes associated to the different paths between transmit and receive antennas. Therefore, the channel impulse response can be given by

$$h(t) = \sum_{i=0}^{I-1} \alpha_i \delta(t - \tau_i), \qquad (A.1)$$

where $\alpha_i$ and $\tau_i$ denote the complex amplitude and the delay associated to the $i$th path, respectively, and $I$ the total number of paths between antennas. The corresponding channel frequency response can then be defined as

$$H(f) = \mathcal{F}\{h(t)\} = \sum_{i=0}^{I-1} \alpha_i \exp\left(-j2\pi f \tau_i\right), \qquad (A.2)$$

and the PDP of the channel can be obtained by

$$\text{PDP}(t) = \sum_{i=0}^{I-1} E\left[|\alpha_i|^2\right] \delta(t - \tau_i), \qquad (A.3)$$

where the complex envelope associated to each path $\alpha_i$ is modeled as a random variable with $\arg(\alpha_i)$ following an uniform distribution in $[0, 2\pi]$ and $|\alpha_i|$ a Rayleigh distribution.

Unless otherwise stated, throughout this work we consider a severely time-dispersive channel characterized by the PDP type C for the HiperLAN/2 [11], shown in Fig. A.1, with uncorrelated Rayleigh fading on the different paths. In some simulations we also consider other severely time-dispersive channel characterized by an exponential PDP, as shown in Fig. A.2, also with uncorrelated Rayleigh fading on the different paths.

199

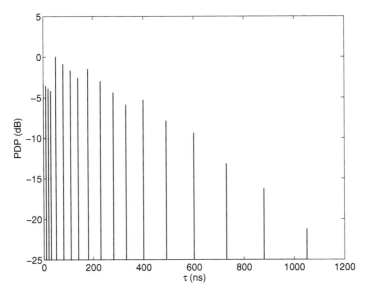

Figure A.1  Reference channel PDP used in this work.

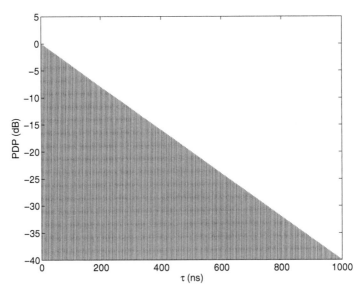

Figure A.2 Exponential PDP channel with uncorrelated Rayleigh fading on the different paths.

# B

# CP and DFT-based Receiver Implementation

Let us consider the received time-domain samples $y_n$ given by the discrete convolution

$$y_n = \sum_{l=0}^{N_h-1} s_{n-l} h_l + w_n, \tag{B.1}$$

where $\{s_n; n = 0, \ldots, N-1\}$ is the block of samples associated to a transmitted burst, $\{h_n; n = 0, \ldots, N_h - 1\}$ is the channel impulse response, with $N_h < N$ denoting the channel length, and $\{w_n; n = 0, \ldots, N-1\}$ the corresponding Gaussian channel noise samples, assumed independent and identically distributed (i.i.d.) in each received burst. Rewriting (B.1) in matrix notation, it is easy to see that it is equivalent to

$$\mathbf{y} = \mathbf{h}\mathbf{s} + \mathbf{w}$$

$$\Leftrightarrow \begin{bmatrix} y_0 \\ y_1 \\ \vdots \\ y_{N-1} \\ -\,-\,- \\ y_N \\ \vdots \\ y_{N+N_h-2} \end{bmatrix}$$

$$
= \begin{bmatrix}
h_0 & 0 & 0 & \cdots & & & & & & & 0 \\
h_1 & h_0 & 0 & 0 & \cdots & & & & & & 0 \\
& \ddots & & \ddots & & & & & & & \\
h_{N_h-1} & h_{N_h-2} & \cdots & h_1 & h_0 & 0 & 0 & \cdots & & & 0 \\
0 & h_{N_h-1} & h_{N_h-2} & \cdots & h_1 & h_0 & 0 & 0 & \cdots & & 0 \\
& & & & \ddots & & \ddots & & & & \\
0 & & & \cdots & & 0 & h_{N_h-1} & h_{N_h-2} & \cdots & h_1 & h_0 \\
\hline
0 & & & \cdots & & & 0 & h_{N_h-1} & h_{N_h-2} & \cdots & h_1 \\
& & & & & & & & \ddots & & \ddots \\
0 & \cdots & & & & & & & 0 & h_{N_h-1} & h_{N_h-2} \\
0 & \cdots & & & & & & & & 0 & h_{N_h-1}
\end{bmatrix}
$$

$$
\times
\begin{bmatrix}
s_0 \\
s_1 \\
\vdots \\
s_{N-1} \\
\hline
s_N \\
\vdots \\
s_{N+N_h-2}
\end{bmatrix}
+
\begin{bmatrix}
w_0 \\
w_1 \\
\vdots \\
w_{N-1} \\
\hline
w_N \\
\vdots \\
w_{N+N_h-2}
\end{bmatrix}.
\tag{B.2}
$$

From (B.2), we can observe that the received time-domain samples associated to the useful part of a given block are

$$
\begin{bmatrix}
y_0 \\
y_1 \\
\vdots \\
y_{N-1}
\end{bmatrix}
=
\overbrace{
\begin{bmatrix}
h_0 & 0 & \cdots & & & & 0 \\
h_1 & h_0 & 0 & \cdots & & & 0 \\
& \ddots & \ddots & & & & \\
0 & \cdots & & 0 & h_{N_h-1} & \cdots & h_1 & h_0 & 0 \\
0 & \cdots & & & 0 & h_{N_h-1} & \cdots & h_1 & h_0
\end{bmatrix}
}^{\mathbf{h_0}_{[N\times N]}}
$$

$$
\times
\begin{bmatrix}
s_0 \\
s_1 \\
\vdots \\
s_{N-1}
\end{bmatrix}
+
\begin{bmatrix}
w_0 \\
w_1 \\
\vdots \\
w_{N-1}
\end{bmatrix}
\tag{B.3}
$$

and the received time-domain samples associated to the interference from the $(m-1)$th block are

$$\overbrace{\qquad\qquad\qquad}^{\mathbf{h}_{1[N\times N]}}$$

$$
\begin{bmatrix} y_N \\ y_{N+1} \\ \vdots \\ y_{2N-1} \end{bmatrix} =
\begin{bmatrix}
0 & \cdots & 0 & h_{N_h-1} & \cdots & h_2 & h_1 \\
0 & \cdots & & 0 & h_{N_h-1} & \cdots & h_2 \\
& & & & \ddots & & \ddots \\
0 & \cdots & & & & 0 & h_{N_h-1} \\
0 & \cdots & & & & & 0 \\
& & & & \vdots & & \\
0 & \cdots & & & & & 0
\end{bmatrix}
$$

$$
\times \begin{bmatrix} s_N \\ s_{N+1} \\ \vdots \\ s_{2N-1} \end{bmatrix} + \begin{bmatrix} w_N \\ w_{N+1} \\ \vdots \\ w_{2N-1} \end{bmatrix}. \tag{B.4}
$$

Then, the received samples associated to the $m$th block can be written as

$$\mathbf{y}^{(m)} = \mathbf{h}_0 \mathbf{s}^{(m)} + \mathbf{h}_1 \mathbf{s}^{(m-1)} + \mathbf{w}^{(m)}. \tag{B.5}$$

Since for cyclic extended bursts with duration $T_B = T_G + T$, at least the first $N_h$ samples associated to the guard period $T_G$ are a repetition of the $L$ final burst samples, this means that $\mathbf{s}^{(m)} = \mathbf{s}^{(m-1)}$ for the guard period, leading to

$$\mathbf{y}^{(m)} = (\mathbf{h}_0 + \mathbf{h}_1)\,\mathbf{s}^{(m)} = \mathbf{h}_{CP}\,\mathbf{s}^{(m)} + \mathbf{w}^{(m)} \tag{B.6}$$

where $\mathbf{h}_{CP}$ is a size-$N \times N$ circulant matrix given by

$$
\mathbf{h}_{CP} = \begin{bmatrix}
h_0 & 0 & \cdots & & 0 & h_{N_h-1} & \cdots & h_2 & h_1 \\
h_1 & h_0 & 0 & \cdots & & 0 & h_{N_h-1} & \cdots & h_2 \\
& \ddots & \ddots & & & & \ddots & & \ddots \\
0 & \cdots & 0 & h_{N_h-1} & \cdots & h_0 & 0 & \cdots & h_{N_h-1} \\
& & & & & & \ddots & & \ddots \\
0 & \cdots & & 0 & h_{N_h-1} & \cdots & h_1 & h_0 & 0 \\
0 & \cdots & & & 0 & h_{N_h-1} & \cdots & h_1 & h_0
\end{bmatrix},
$$

$$\tag{B.7}$$

which elements verify $[\mathbf{h}_{CP}]_{i,i'} = h_{(i-i')\bmod N}$ ($x \bmod y$ denotes the modulo operation, i.e., the remainder of division of $x$ by $y$).

It is well known that circulant matrices can be diagonallized using a Fourier matrix, i.e.,

$$\mathbf{h}_{CP} = \mathfrak{F}^{-1} \mathbf{\Lambda} \mathfrak{F}, \tag{B.8}$$

where the size-$N \times N$ matrix $\mathfrak{F}$, given by

$$\mathfrak{F} = \frac{1}{\sqrt{N}} \begin{bmatrix} 1 & 1 & 1 & \cdots & 1 \\ 1 & \omega & \omega^2 & \cdots & \omega^{N-1} \\ 1 & \omega^2 & \omega^4 & \cdots & \omega^{2(N-1)} \\ \vdots & \vdots & \vdots & & \vdots \\ 1 & \omega^{N-1} & \omega^{2(N-1)} & \vdots & \omega^{(N-1)(N-1)} \end{bmatrix}, \qquad (B.9)$$

with $\omega = \exp(-j2\pi/N)$, is the unitary (i.e., $\mathfrak{F}^H = \mathfrak{F}^{-1}$) DFT matrix whose columns are the eigenvectors of $\mathbf{h}_{CP}$, and $\mathbf{\Lambda}$ is a size-$N \times N$ diagonal matrix whose elements, the eigenvalues of $\mathbf{h}_{CP}$, equals the DFT of the first column of $\mathbf{h}_{CP}$, i.e.,

$$\mathbf{\Lambda} = \begin{bmatrix} \lambda_0 & & & 0 \\ & \lambda_1 & & \\ & & \ddots & \\ 0 & & & \lambda_{N-1} \end{bmatrix} = \begin{bmatrix} H_0 & & & 0 \\ & H_1 & & \\ & & \ddots & \\ 0 & & & H_{N-1} \end{bmatrix} = \mathbf{H}, \quad (B.10)$$

where

$$\lambda_k = \sum_{n=0}^{N-1} h_n \exp\left(-j2\pi \frac{kn}{N}\right) = \mathrm{DFT}\{h_n\} = H_k, \quad k = 0, 1, \ldots, N-1.$$

$$(B.11)$$

Using (B.8) and (B.10) in (B.6), it follows that (for the sake of notation simplicity we will drop the superscript $m$)

$$\mathbf{y} = \mathfrak{F}^{-1}\mathbf{H}\mathfrak{F}\mathbf{s} + \mathbf{w} \Leftrightarrow \mathfrak{F}\mathbf{y} = \mathbf{H}\mathfrak{F}\mathbf{s} + \mathfrak{F}\mathbf{w}$$
$$\Leftrightarrow \mathbf{Y} = \mathbf{H}\mathbf{S} + \mathbf{N}, \qquad (B.12)$$

with the vectors

$$\mathbf{Y} = \mathfrak{F}\mathbf{y} = \begin{bmatrix} Y_0 \\ Y_1 \\ \vdots \\ Y_{N-1} \end{bmatrix}, \qquad (B.13)$$

$$\mathbf{S} = \mathfrak{F}\mathbf{s} = \begin{bmatrix} S_0 \\ S_1 \\ \vdots \\ S_{N-1} \end{bmatrix}, \qquad (B.14)$$

and

$$\mathbf{N} = \mathfrak{F}\mathbf{w} = \begin{bmatrix} N_0 \\ N_1 \\ \vdots \\ N_{N-1} \end{bmatrix} \tag{B.15}$$

denoting the DFTs of $\mathbf{y}$, $\mathbf{s}$ and $\mathbf{w}$, respectively. This means that the received sample at the $k$th subcarrier is given by

$$Y_k = H_k S_k + N_k, \tag{B.16}$$

$k = 0, 1, \ldots, N - 1$, where $H_k$ is the channel frequency response for the $k$th subcarrier and $N_k$ the channel noise component for that subcarrier.

# C

---

# Computation of LLR for QPSK Constellation

---

Let us consider that the frequency-domain symbols associated to the $p$th user $\{A_{m,p}; m = 0, \ldots, M-1\}$ in an MC-CDMA transmission are selected from a QPSK constellation. The frequency-domain samples $\{\tilde{A}_{m,p}; m = 0, \ldots, M-1\}$ at the FDE output can be written as

$$\tilde{A}_{m,p} = A_{m,p} + \Theta_{m,p} = A^I_{m,p} + jA^Q_{m,p} + \Theta_{m,p} \tag{C.1}$$

where $A^I_{m,p}$ ($A^Q_{m,p}$) denotes the "in-phase" ("quadrature") bit of the $m$th transmitted symbol and $\Theta_{m,p}$ denotes the corresponding frequency-domain overall error component associated to the $p$th user. Moreover, let us assume that $\Theta_{m,p}$ is Gaussian-distributed, with mean $E[\Theta_{m,p}] = 0$ and variance

$$2\sigma_p^2 = E\left[\left|A_{m,p} - \tilde{A}_{m,p}\right|^2\right], \tag{C.2}$$

and that $A^I_{m,p}$ and $A^Q_{m,p}$ are equiprobable, i.e.,

$$\Pr(A^I_{m,p} = \pm 1) = \Pr(A^Q_{m,p} = \pm 1) = \frac{1}{2}. \tag{C.3}$$

The LLRs associated to the "in-phase" and "quadrature" bits of the $m$th transmitted symbol are defined, respectively, as

$$L^I_{m,p} = \ln\left(\frac{\Pr\left(A^I_{m,p} = 1 | \tilde{A}^I_{m,p}\right)}{\Pr\left(A^I_{m,p} = -1 | \tilde{A}^I_{m,p}\right)}\right) \tag{C.4}$$

and

$$L^Q_{m,p} = \ln\left(\frac{\Pr\left(A^Q_{m,p} = 1 | \tilde{A}^Q_{m,p}\right)}{\Pr\left(A^Q_{m,p} = -1 | \tilde{A}^Q_{m,p}\right)}\right). \tag{C.5}$$

By applying Bayes' rule, one obtains

$$
\Pr\left(A^I_{m,p} = \pm 1 | \tilde{A}^I_{m,p}\right) = \frac{p_A\left(\tilde{A}^I_{m,p} | A^I_{m,p} = \pm 1\right) \Pr\left(A^I_{m,p} = \pm 1\right)}{p_A\left(\tilde{A}^I_{m,p}\right)}
$$

$$
= \frac{p_A\left(\tilde{A}^I_{m,p} | A^I_{m,p} = \pm 1\right)}{2\, p_A\left(\tilde{A}^I_{m,p}\right)} \tag{C.6}
$$

where $p_A(\tilde{A}^I_{m,p})$ and $p_A(\tilde{A}^I_{m,p} | A^I_{m,p} = \pm 1)$ denote the pdf and the conditional pdf of $\tilde{A}^I_{m,p}$, respectively, given by

$$
p_A\left(\tilde{A}^I_{m,p}\right) = p_A\left(\tilde{A}^I_{m,p} | A^I_{m,p} = 1\right) \Pr\left(A^I_{m,p} = 1\right)
$$
$$
+ p_A\left(\tilde{A}^I_{m,p} | A^I_{m,p} = -1\right) \Pr\left(A^I_{m,p} = -1\right)
$$
$$
= \frac{1}{2}\, p_A\left(\tilde{A}^I_{m,p} | A^I_{m,p} = 1\right) + \frac{1}{2}\, p_A\left(\tilde{A}^I_{m,p} | A^I_{m,p} = -1\right) \tag{C.7}
$$

and

$$
p_A\left(\tilde{A}^I_{m,p} | A^I_{m,p} = \pm 1\right) = \frac{1}{\sqrt{2\pi\sigma_p^2}} \exp\left(-\frac{\left(\tilde{A}^I_{m,p} \mp 1\right)^2}{2\sigma_p^2}\right). \tag{C.8}
$$

Then (C.7) can be expressed as

$$
p_A\left(\tilde{A}^I_{m,p}\right) = \frac{1}{2\sqrt{2\pi\sigma_p^2}}\left(\exp\left(-\frac{\left(\tilde{A}^I_{m,p} - 1\right)^2}{2\sigma_p^2}\right)\right.
$$
$$
\left. + \exp\left(-\frac{\left(\tilde{A}^I_{m,p} + 1\right)^2}{2\sigma_p^2}\right)\right). \tag{C.9}
$$

Substituting (C.8) and (C.9) in (C.6), results in

$$\Pr\left(\tilde{A}^I_{m,p} = \pm 1 | \tilde{A}^I_{m,p}\right)$$

$$= \frac{\exp\left(-\frac{\left(\tilde{A}^I_{m,p} \mp 1\right)^2}{2\sigma_p^2}\right)}{\exp\left(-\frac{\left(\tilde{A}^I_{m,p} - 1\right)^2}{2\sigma_p^2}\right) + \exp\left(-\frac{\left(\tilde{A}^I_{m,p} + 1\right)^2}{2\sigma_p^2}\right)}. \quad \text{(C.10)}$$

This means that (C.4) can be written as

$$L^I_{m,p} = \frac{\left(\tilde{A}^I_{m,p} + 1\right)^2}{2\sigma_p^2} - \frac{\left(\tilde{A}^I_{m,p} - 1\right)^2}{2\sigma_p^2} = \frac{2}{\sigma_p^2}\tilde{A}^I_{m,p}. \quad \text{(C.11)}$$

Similarly, (C.5) can be written as

$$L^Q_{m,p} = \frac{2}{\sigma_p^2}\tilde{A}^Q_{m,p}. \quad \text{(C.12)}$$

The mean value of $A_{m,p}$, i.e., the "soft decision" $\overline{A}_{m,p}$, is given by

$$\overline{A}_{m,p} = E\left[A_{m,p} | \tilde{A}_{m,p}\right] = E\left[A^I_{m,p} | \tilde{A}^I_{m,p}\right] + jE\left[A^Q_{m,p} | \tilde{A}^Q_{m,p}\right], \quad \text{(C.13)}$$

where

$$E\left[A^I_{m,p} | \tilde{A}^I_{m,p}\right] = \Pr\left(A^I_{m,p} = 1 | \tilde{A}^I_{m,p}\right) - \Pr\left(A^I_{m,p} = -1 | \tilde{A}^I_{m,p}\right)$$

$$= \frac{p_A\left(\tilde{A}^I_{m,p} | A^I_{m,p} = 1\right)}{2\, p_A\left(\tilde{A}^I_{m,p}\right)} - \frac{p_A\left(\tilde{A}^I_{m,p} | A^I_{m,p} = -1\right)}{2\, p_A\left(\tilde{A}^I_{m,p}\right)}$$

$$= \frac{\exp\left(-\frac{\left(\tilde{A}^I_{m,p} - 1\right)^2}{2\sigma_p^2}\right) - \exp\left(-\frac{\left(\tilde{A}^I_{m,p} + 1\right)^2}{2\sigma_p^2}\right)}{\exp\left(-\frac{\left(\tilde{A}^I_{m,p} - 1\right)^2}{2\sigma_p^2}\right) + \exp\left(-\frac{\left(\tilde{A}^I_{m,p} + 1\right)^2}{2\sigma_p^2}\right)}$$

$$= \frac{\exp\left(\dfrac{\tilde{A}^I_{m,p}}{\sigma^2_p}\right) - \exp\left(-\dfrac{\tilde{A}^I_{m,p}}{\sigma^2_p}\right)}{\exp\left(\dfrac{\tilde{A}^I_{m,p}}{\sigma^2_p}\right) + \exp\left(-\dfrac{\tilde{A}^I_{m,p}}{2\sigma^2_p}\right)}$$

$$= \tanh\left(\frac{\tilde{A}^I_{m,p}}{\sigma^2_p}\right) = \tanh\left(\frac{L^I_{m,p}}{2}\right) \tag{C.14}$$

and

$$E\left[A^Q_{m,p} | \tilde{A}^Q_{m,p}\right] = \tanh\left(\frac{L^Q_{m,p}}{2}\right). \tag{C.15}$$

Therefore, the "soft decision" value can be written as a function of the LLRs (C.11) and (C.20) by

$$\overline{A}_{m,p} = \tanh\left(\frac{L^I_{m,p}}{2}\right) + j \tanh\left(\frac{L^Q_{m,p}}{2}\right). \tag{C.16}$$

For a DS-CDMA transmission the transmitted symbols associated to the $p$th user are $\{a_{m,p}; m = 0, \ldots, M - 1\}$ and the time-domain samples are $\{\tilde{a}_{m,p}; m = 0, \ldots, M - 1\}$, with

$$\tilde{a}_{m,p} = a_{m,p} + \theta_{m,p} = a^I_{m,p} + ja^Q_{m,p} + \theta_{m,p}, \tag{C.17}$$

where $a^I_{m,p}$ ($a^Q_{m,p}$) denotes the "in-phase" ("quadrature") bit of the $m$th transmitted symbol and $\theta_{m,p}$ denotes the corresponding Gaussian error component with mean $E[\theta_{m,p}] = 0$ and variance

$$\sigma^2_p = \frac{1}{2} E\left[|a_{m,p} - \tilde{a}_{m,p}|^2\right]. \tag{C.18}$$

Following the same reasoning as for the MC-CDMA case described above, it can easily be shown that for DS-CDMA the LLRs associated to the "in-phase" and "quadrature" bit of $a_{m,p}$ are

$$L^I_{m,p} = \frac{2}{\sigma^2_p} \tilde{a}^I_{m,p} \tag{C.19}$$

and

$$L^Q_{m,p} = \frac{2}{\sigma^2_p} \tilde{a}^Q_{m,p}, \tag{C.20}$$

respectively. In this case, the "soft decision" value of $a_{m,p}$ is

$$\bar{a}_{m,p} = \tanh\left(\frac{L^I_{m,p}}{2}\right) + j \tanh\left(\frac{L^Q_{m,p}}{2}\right). \tag{C.21}$$

# D

# Equivalence between (4.32) and (4.40)

In the following we will show that the optimum feedforward coefficient vector $\mathbf{F}_p(k)$ given by (4.32) can be written as (4.40) for a SIC receiver (the extension to PIC receivers is straightforward).

Let us rewrite (4.32) and (4.40) using non-matricial notation

$$\sum_{p'=1}^{P} \left(1 - \rho_{p'}^2\right) H_{k,p'}^* H_{k,p'} F_{k,p} + \beta F_{k,p} = Q_p H_{k,p}^*, \tag{D.1}$$

where $Q_p = \gamma_p(1 - \rho_p^2) - \lambda_p/(2\sigma_A^2 M)$, and

$$F_{k,p} = \sum_{p'=1}^{P} H_{k,p'}^* [\mathbf{V}(k)]_{p,p'} [\mathbf{Q}]_{p,p'}, \tag{D.2}$$

respectively ($[\mathbf{A}]_{n,m}$ denote the element of line $n$ and column $m$ of matrix $\mathbf{A}$). Substituting (D.2) in (D.1) it follows that

$$\sum_{p'=1}^{P}(1 - \rho_{p'}^2)H_{k,p'}^* H_{k,p'} \sum_{p''=1}^{P} H_{k,p''}^* [\mathbf{G}(k)]_{p,p''}[\mathbf{Q}]_{p,p''}$$

$$+\beta \sum_{p''=1}^{P} H_{k,p''}^* [\mathbf{V}(k)]_{p,p''}[\mathbf{Q}]_{p,p''} = Q_p H_{k,p}^*$$

$$\Leftrightarrow \sum_{p'=1}^{P}(1 - \rho_{p'}^2)H_{k,p'}^* H_{k,p'} \sum_{p''=1}^{P} H_{k,p''}^* [\mathbf{V}(k)]_{p,p''}[\mathbf{Q}]_{p,p''}$$

$$+\beta \sum_{p'=1}^{P}\sum_{p''=1}^{P} H_{k,p''}^* [\mathbf{V}(k)]_{p,p''}[\mathbf{Q}]_{p,p''}\delta_{p',p''} = Q_p H_{k,p}^*$$

$$\Leftrightarrow \sum_{p'=1}^{P} H_{k,p'}^{*} \left[ \sum_{p''=1}^{P} [\mathbf{V}(k)]_{p,p''} [\mathbf{Q}]_{p,p''} \left( \left(1 - \rho_{p'}^{2}\right) H_{k,p''}^{*} H_{k,p'} + \beta \delta_{p',p''} \right) \right]$$
$$= Q_{p} H_{k,p}^{*} \tag{D.3}$$

($\delta_{p,p'} = 1$ if $p = p'$ and 0 otherwise). From (4.41), which, in non-matricial notation, is given by

$$\sum_{p''=1}^{P} [\mathbf{V}(k)]_{p,p''} \left( \left(1 - \rho_{p'}^{2}\right) H_{k,p''}^{*} H_{k,p'} + \beta \delta_{p',p''} \right) = \delta_{p,p'}, \tag{D.4}$$

we can easily see that the factor between brackets in (D.3) reduces to

$$\sum_{p''=1}^{P} [\mathbf{V}(k)]_{p,p''} [\mathbf{Q}]_{p,p''} \left( \left(1 - \rho_{p'}^{2}\right) H_{k,p''}^{*} H_{k,p'} + \beta \delta_{p',p''} \right) = [\mathbf{Q}]_{p,p'} \delta_{p,p'}.$$
$$\tag{D.5}$$

Substituting the result above in (D.3) leads to

$$\sum_{p'=1}^{P} H_{k,p'}^{*} [\mathbf{Q}]_{p,p'} \delta_{p,p'} = Q_{p} H_{k,p}^{*} \tag{D.6}$$

which completes the demonstration.

# E

# Power Amplification

The power amplification of bandpass signals usually employs amplifiers that can be modeled as bandpass memoryless nonlinearities [84]. For instance, a solid state power amplifier (SSPA) is characterized with AM-to-AM and AM-to-PM conversions given by

$$A(R) = \frac{R\,\dfrac{A_M}{s_M}}{\sqrt[2p]{1 + \left(\dfrac{R}{s_M}\right)^{2p}}} \tag{E.1}$$

and

$$\Theta(R) \approx 0, \tag{E.2}$$

respectively, where $A_M$ denotes the maximum value of the output envelope (i.e., $A_M = \lim_{R \to +\infty} A(R)$), $A_M/s_M$ is the low power amplification factor (i.e., $A_M/s_M = \lim_{R \to 0} A(R)/R$) and $p$ is the parameter that controls the transition softness between the linear and saturation regions, as shown in Fig. E.1.

Since $\theta(R) \approx 0$ and $A(R) \approx R\,A_M/s_M$ almost until the saturation region, then for large values of $p$ SSPA allow a low nonlinear distortion, avoiding the necessity of pre-distortion techniques for linearization proposes (note that, for $p = +\infty$, (E.1) corresponds to an ideal hard-limiter). This is the reason why SSPAs have been used in several wireless communications systems with strong envelope fluctuations signals, namely with multicarrier modulations, despite there small efficiency and low maximum output power.

With traveling wave tube amplifier (TWTA) the amplification efficiency as well as maximum output power is higher. Their AM-to-AM and AM-to-

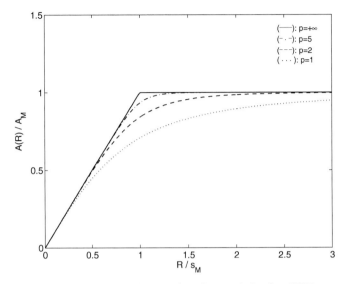

Figure E.1  AM-to-AM conversion characteristic of an SSPA.

PM conversions characteristics are given by

$$A(R) = 2\frac{R\,\dfrac{A_M}{s_M}}{1 + \left(\dfrac{R}{s_M}\right)^2} \tag{E.3}$$

and

$$\Theta(R) = 2\frac{\theta_M \left(\dfrac{R}{s_M}\right)^2}{1 + \left(\dfrac{R}{s_M}\right)^2}, \tag{E.4}$$

respectively, where $s_M$ and $A_M$ denote the input and output envelope values at saturation, respectively, and $\theta_M$ is the corresponding phase rotation [84] (see Fig. E.2).

In general, power amplifiers with high efficiency and high output power are strongly nonlinear. This type of amplifier are usually modeled as an ideal hard-limiter, corresponding to a bandpass memoryless nonlinearity where

$$A(R) = A_M \tag{E.5}$$

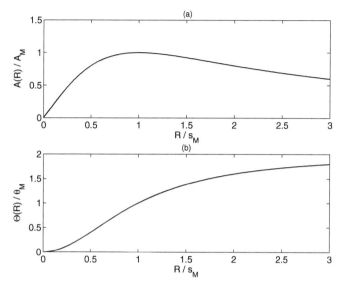

Figure E.2 (a) AM-to-AM and (b) AM-to-PM conversion characteristics of an TWTA with $s_M = A_M = 1$.

and

$$\Theta(R) = 0 \qquad\qquad (E.6)$$

with $A_M$ denoting the output envelope. In this case, the nonlinear distortion effects are very strong unless input signals have quasi-constant envelope.

# F

---

## Nonlinear Effects on Gaussian Signals

---

This appendix presents some results concerning to the impact of memoryless bandpass nonlinearities on Gaussian signals.

Let us consider a bandpass signal, $x_{BP}(t)$, described by

$$
\begin{aligned}
x_{BP}(t) &= \frac{1}{2}x_{in}(t)\exp(j2\pi f_c t) + \frac{1}{2}x_{in}^*(t)\exp(-j2\pi f_c t) \\
&= \mathrm{Re}\left\{x_{in}(t)\exp(j2\pi f_c t)\right\}
\end{aligned}
\tag{F.1}
$$

where $f_c$ is the carrier frequency and

$$
x_{in}(t) = R(t)\exp(j\varphi(t))
\tag{F.2}
$$

is the complex envelope of $x_{BP}(t)$, with $R(t) = |x_{in}(t)|$ and $\varphi(t) = \arg(x_{in}(t))$ representing the envelope and phase of $x_{in}(t)$, respectively. Clearly,

$$
x_{BP}(t) = \mathrm{Re}\{R(t)\exp(j\psi(t))\} = R(t)\cos(\psi(t))
\tag{F.3}
$$

with

$$
\psi(t) = \varphi(t) + 2\pi f_c t.
\tag{F.4}
$$

Let us suppose now that the signal $x_{BP}(t)$ is submitted to the memoryless bandpass nonlinearity represented in Fig. F.1. It can be shown that the complex envelope of the signal at the nonlinearity output is given by

$$
y_{out}(t) = A(R)\exp(j\Theta(R))\exp(j\varphi),
\tag{F.5}
$$

where $A(R)$ and $\Theta(R)$ represent, respectively, the AM-to-AM and AM-to-PM conversion functions of the nonlinearity given by

$$
A(R)\exp(j\Theta(R)) = \begin{cases} R, & R \le s_M \\ s_M, & R > s_M \end{cases},
\tag{F.6}
$$

where $s_M$ corresponds to the envelope clipping level, and the AM-to-PM conversion function $\Theta(R) = 0$.

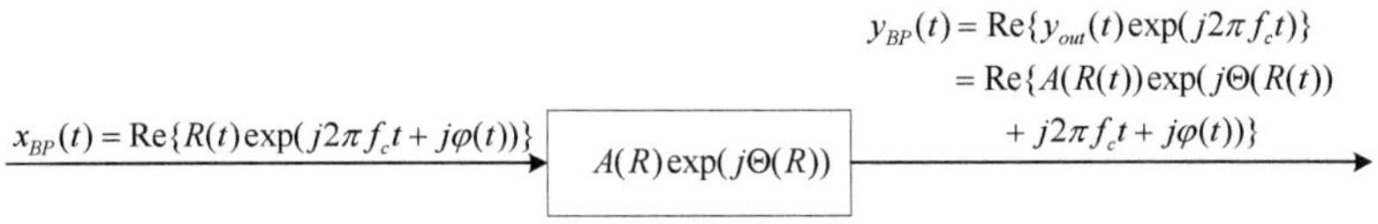

$$y_{BP}(t) = \text{Re}\{y_{out}(t)\exp(j2\pi f_c t)\}$$
$$= \text{Re}\{A(R(t))\exp(j\Theta(R(t))$$

$$x_{BP}(t) = \text{Re}\{R(t)\exp(j2\pi f_c t + j\varphi(t))\}$$

$$A(R)\exp(j\Theta(R))$$

$$+ j2\pi f_c t + j\varphi(t))\}$$

Figure F.1 Bandpass memoryless nonlinearity characterization by their AM-to-AM and AM-to-PM conversion functions.

It can be shown [97] that the autocorrelation function of $y_{out}(t)$ is given by

$$R_{out}(\tau) = 2 \sum_{\gamma=0}^{+\infty} P_{2\gamma+1} \frac{(R_{in}(\tau))^{\gamma+1} \left(R_{in}^*(\tau)\right)^{\gamma}}{(R_{in}(0))^{2\gamma+1}} \qquad (F.7)$$

where $R_{in}(\tau) = E[x_{in}(t)x_{in}^*(t-\tau)]$ represent the autocorrelation of $x_{in}(t)$ and $P_{2\gamma+1}$ is the power associated the inter-modulation product (IMP) of order $2\gamma + 1$. This power can be calculated as the approach given in [96], and is given by

$$P_{2\gamma+1} = \frac{\left|\vartheta_{2\gamma+1}\right|^2}{2\gamma!(\gamma+1)!} \qquad (F.8)$$

where

$$\vartheta_{2\gamma+1} = \frac{2}{\sigma^2} \int_0^{+\infty} RA(R)\exp(j\Theta(R))W_{2\gamma+1}\left(\frac{R}{\sqrt{2\sigma^2}}\right) dR \qquad (F.9)$$

with

$$W_{2\gamma+1}(x) = \frac{\gamma!}{2} \exp\left(-x^2\right) x L_\gamma^{(1)}\left(x^2\right), \qquad (F.10)$$

where $L_\gamma^{(1)}(x)$ denotes a generalized Laguerre polynomial of order $\gamma$ [102] defined as

$$L_\gamma^{(1)}(x) = \frac{1}{\gamma!x} \exp(x) \frac{d^\gamma}{dx^\gamma} \left\{\exp(-x)x^{\gamma+1}\right\}. \qquad (F.11)$$

From (F.8) to (F.11) it follows that $P_1 = |\alpha|^2\sigma^2$, with $\alpha$ given by

$$\alpha = \frac{1}{2\sigma^2} \int_0^{+\infty} RA(R)\exp(j\Theta(R)) \frac{R}{\sigma^2} \exp\left(-\frac{R^2}{2\sigma^2}\right) dR. \qquad (F.12)$$

Moreover, since the envelope $R = |x_{in}(t)|$ has a Rayleigh distribution, $\alpha$ can be written as

$$\alpha = \frac{E\left[RA(R)\exp(j\Theta(R))\right]}{E\left[R^2\right]}. \qquad (F.13)$$

This means that, apart from a scalar factor $\alpha$, the first IMP corresponds to the input signal; on the other hand, since the signal component associated to the remaining IMPs is uncorrelated to the input signal, we can write

$$y_{out}(t) = \alpha x_{in}(t) + d(t), \tag{F.14}$$

where $d(t)$ represent the complex envelope of a self-interference component, which is uncorrelated with $x_{in}(t)$.

The power of the self-interference component is given by

$$I = \sum_{\gamma=1}^{+\infty} P_{2\gamma+1} = P_{out} - S \tag{F.15}$$

where $S = P_1 = |\alpha|^2 \sigma^2$ is the power of the useful component and $P_{out}$ the total power at the nonlinearity output, given by

$$P_{out} = \frac{1}{2} E\left[A^2(R)\right] = \frac{1}{2} \int_0^{+\infty} A^2(R) \frac{R}{\sigma^2} \exp\left(-\frac{R^2}{2\sigma^2}\right) dR \tag{F.16}$$

Since the useful and self-interference components are uncorrelated, we can write

$$R_{out}(\tau) = |\alpha|^2 R_{in}(\tau) + R_d(\tau), \tag{F.17}$$

where $R_d(\tau) = E[d(t)d^*(t-\tau)]$ represents the autocorrelation of the self-interference component, given by

$$R_d(\tau) = 2 \sum_{\gamma=1}^{+\infty} P_{2\gamma+1} \left(\frac{R_{in}(\tau)}{R_{in}(0)}\right)^{\gamma+1} \left(\frac{R_{in}^*(\tau)}{R_{in}(0)}\right)^{\gamma}, \tag{F.18}$$

with $R_d(0) = I = \sum_{\gamma=1}^{+\infty} P_{2\gamma+1}$. Moreover, the power spectral density associated with the self-interference component is given by

$$G_d(f) = \mathcal{F}\{R_d(\tau)\} = 2 \sum_{\gamma=1}^{+\infty} P_{2\gamma+1} \mathcal{F}\left\{\frac{(R_{in}(\tau))^{\gamma+1} (R_{in}^*(\tau))^{\gamma}}{(R_{in}(0))^{2\gamma+1}}\right\}$$

$$= 2 \sum_{\gamma=1}^{+\infty} \frac{P_{2\gamma+1}}{(R_{in}(0))^{2\gamma+1}} G_{in}(f)$$

$$* \underbrace{(G_{in}(f) * G_{in}(-f)) * \ldots * (G_{in}(f) * G_{in}(-f))}_{\gamma \text{ times}}$$

$$\tag{F.19}$$

# Bibliography

[1] G. Forney. Maximum-likelihood sequence estimation of digital sequences in the presence of intersymbol interference. *IEEE Trans. Inf. Theory*, 18(3):363–378, May 1972.

[2] J. Proakis. *Digital Communications*. 4th Edition, McGraw-Hill, 2001.

[3] L. Cimini Jr. Analysis and simulation of a digital mobile channel using orthogonal frequency division multiplexing. *IEEE Trans. Commun.*, 33(7):665–675, July 1985.

[4] J. A. C. Bingham. Multicarrier modulation for data transmission: An idea whose time has come. *IEEE Trans. Commun.*, 28(5):5–14, May 1990.

[5] R. van Nee and R. Prasad. *OFDM for Wireless Multimedia Communications*. Artech House Publ., 2000.

[6] J. W. Cooley and J. W. Tukey. An algorithm for the machine calculation of complex fourier series. *Mathematics of Computation*, 19(90):297–301, Apr. 1965.

[7] T. Ojamperä and R. Prasad. *Wideband CDMA for Third Generation Mobile Communications*. Artech House Publ., 1998.

[8] A. Viterbi. *CDMA: Principles of SS Communication*. Addison Wesley, 1995.

[9] S. Hara and R. Prasad. Overview of multicarrier CDMA. *IEEE Commun. Magazine*, 35(12):126–133, Dec. 1997.

[10] N. Benvenuto and S. Tomasin. Block iterative DFE for single carrier modulation. *IEE Electron. Lett.*, 39(19):1144–1145, Sep. 2002.

[11] ETSI. Channel models for HiperLAN/2 in different indoor scenarios. *ETSI EP BRAN 3ERI085B*, pages 1–8, Mar. 1998.

[12] A. Bruce Carlson. *Communications Systems: An Introduction to Signals and Noise in Electrical Communication*, pages 622–627. 3rd Edition, McGraw-Hill, New York, 1986.

[13] B. Muquet, M. Courville, P. Dunamel, and G. Giannakis. OFDM with trailing zeros versus OFDM with cyclic prefix: Links, comparisons and application to the HiperLAN/2 system. In *IEEE ICC'00*, volume 2, pages 1049–1053, Santa-Barbara, California, June 2000.

[14] B. Muquet, M. Courville, G. Giannakis, Z. Wang, and P. Dunamel. Reduced complexity equalizers for zero-padded OFDM transmission. In *IEEE ICASSP'00*, volume 5, pages 2973–2976, Istanbul, Turkey, June 2000.

[15] L. Deneire, B. Gyselinckx, and M. Engels. Training sequence versus cyclic prefix - a new look on single carrier communications. *IEEE Commun. Lett.*, 5(7):292–294, July 2001.

[16] T. Araújo and R. Dinis. Efficient detection of zero-padded OFDM signals with large blocks. In *IASTED SIP'06*, Honolulu, Hawaii, Aug. 2006.

[17] A. Stamoulis, G. B. Giannakis, and A. Scaglioni. Block FIR decision-feedback equalizer for filterbank precoded transmissions with blind estimation capabilities. *IEEE Trans. Commun.*, 49(1):69–83, Jan. 2001.

[18] A. Scaglione, G. B. Giannakis, and S. Barbarossa. Redundant filterbank precoders and equalizers - Part I: Unification and optimal designs, and Part II: Blind channel estimation, synchronization and direct equalization. *IEEE Trans. Signal Process.*, 47(7):1988–2022, July 1999.

[19] N. Benvenuto and G. Cherubini. *Algorithms for Communication Systems and Their Applications*. Chichester, UK: Wiley, 2002.

[20] D. Falconer and S. Ariyavisitakul. Broadband wireless using single carrier and frequency domain equalization. In *IEEE WCNC*, volume 1, pages 27–36, Honolulu, USA, Oct. 2002.

[21] S. Kaiser. On the performance of different detection techniques for OFDM-CDMA in fading channels. In *IEEE GLOBECOM'95*, volume 3, pages 2059–2063, Singapore, Nov. 1995.

[22] A. Gusmão, R. Dinis, J. Conceição, and N. Esteves. Comparison of two modulation choices for broadband wireless communications. In *IEEE VTC'00 (Spring)*, volume 2, pages 1300–1305, Tokyo, Japan, May 2000.

[23] D. Falconer, S. Ariyavisitakul, A. Benyamin-Seeyar, and B. Eidson. Frequency domain equalization for single-carrier broadband wireless systems. *IEEE Commun. Mag.*, 4(4):58–66, Apr. 2002.

[24] H. Sari, G. Karam, and I. Jeanclaude. Frequency-domain equalization of mobile radio and terrestrial broadcast channels. In *IEEE GLOBECOM'94*, volume 1, pages 1–5, San Francisco, CA, Nov. 1994.

[25] H. Sari, G. Karam, and I. Jeanclaude. Transmission techniques for digital terrestrial TV broadcasting. *IEEE Trans. Commun.*, 33(2):100–109, Feb. 1995.

[26] A. Gusmão, R. Dinis, and N. Esteves. On frequency-domain equalization and diversity combining for broadband wireless communications. *IEEE Trans. Commun.*, 51(7):1029–1033, July 2003.

[27] H. Sari, G. Karam, and I. Jeanclaude. An analysis of orthogonal frequency-division multiplexing for mobile radio applications. In *IEEE VTC'94*, volume 3, pages 1635–1639, Stockholm, June 1994.

[28] S. Müller and J. Huber. A comparison of peak reduction schemes for OFDM. In *IEEE GLOBECOM'97*, volume 1, pages 1–5, Phoenix, Arizona, Nov. 1997.

[29] X. Li and L. Cimini. Effects of clipping and filtering on the performance of OFDM. *IEEE Commun. Lett.*, 2(5):131–133, May 1998.

[30] R. Dinis and A. Gusmão. A class of nonlinear signal processing schemes for bandwidth-efficient OFDM transmission with low envelope fluctuation. *IEEE Trans. Commun.*, 52(11):2009–2018, Nov. 2004.

[31] P. Silva and R. Dinis. A technique for reducing the PMEPR of MC-CDMA signals. In *ECWT'04*, pages 25–28, Amsterdam, Oct. 2004.

[32] R. Dinis and P. Silva. Analytical evaluation of nonlinear effects in MC-CDMA signals. *Trans. Wir. Commun.*, 5(8):2277–2284, Aug. 2006.

[33] N. Benvenuto and S. Tomasin. On the comparison between OFDM and single carrier modulation with a DFE using a frequency domain feedforward filter. *IEEE Trans. Commun.*, 50(6):947–955, Jun. 2002.

[34] R. Dinis, A. Gusmão, and N. Esteves. On broadband block transmission over strongly frequency-selective fading channels. In *15th International Conference on Wireless Communications (Wireless 2003)*, pages 261–269, Calgary, Canada, July 2003.

[35] R. Dinis, A. Gusmão, and N. Esteves. Iterative block-DFE techniques for single-carrier-based broadband communications with transmit/receive space diversity. In *IEEE ISWCS'04*, pages 347–351, Mauritius, Sep. 2004.

[36] M. Tüchler, R. Koetter, and A. Singer. Turbo equalization: Principles and new results. *IEEE Trans. Commun.*, 50(5):754–767, May 2002.

[37] S. Hara and R. Prasad. Design and performance of multicarrier CDMA system in frequency-selective rayleigh fading channels. *IEEE Trans. Veh. Technol.*, 48(5):1584–1595, Sep. 1999.

[38] N. Yee, J-P. Linnartz, and G. Fettweis. Multicarrier CDMA for indoor wireless radio networks. In *IEEE PIMRC'93*, pages 109–113, Yokohama, Japan, Sep. 1993.

[39] K. Fazel and L. Papke. On the performance of convolutional-coded CDMA/OFDM for mobile communication systems. In *IEEE PIMRC'93*, pages 468–472, Yokohama, Japan, Sep. 1993.

[40] A. Chouly, A. Brajal, and S. Jourdan. Orthogonal multicarrier techniques applied to direct sequence spread spectrum CDMA. In *IEEE GLOBECOM'93*, volume 3, pages 1723–1728, Huston, USA, Nov. 1993.

[41] V. DaSilva and E. S. Sousa. Peformance of orthogonal CDMA codes for quasi-synchronous communications systems. In *IEEE ICUPC'93*, pages 995–999, Ottawa, Canada, Oct. 1993.

[42] L. Vandendorpe. Multitone direct sequence CDMA systems in an indoor wireless environment. In *IEEE First Symp. on Commun. & Veh. Technol. in the Benelux*, pages 4.1.1–4.1.8, Delft, Netherlands, Oct. 1993.

[43] H. Sari. Orthogonal multicarrier CDMA and its detection on frequency-selective channels. *Eur. Trans. Telecommun. (ETT), John Wiley & Sons*, 13(5):439–445, Sep.–Oct. 2002.

[44] R. Le Gouable. *Association de Modulations Multiporteuses et de Techniques d'Acces Multiple: Applications aux Transmissions Sans Fil à Haut Débit*. PhD thesis, INSA, Rennes, Mars 2001.

[45] W. Gardner. Exploitation of spectral redundancy in cyclostationary signals. *IEEE Signal Processing Mag.*, 8(2):14–36, Aug. 1991.

[46] S. Kaiser and J. Hagenauer. Multi-carrier CDMA with iterative decoding and soft-interference cancelation. In *IEEE GLOBECOM'97*, volume 1, pages 6–10, Phoenix, Arizona, Nov. 1997.

[47] X. Wang and H. Poor. Iterative (turbo) soft interference cancelation and decoding for coded CDMA. *IEEE Trans. Commun.*, 47(7):1046–1061, July 1999.

[48] P. Silva and R. Dinis. An iterative frequency-domain decision feedback receiver for CDMA systems. In *IEEE ISWCS'04*, pages 6–10, Mauritius, Sep. 2004.

[49] R. Dinis, P. Silva, and A. Gusmão. An iterative frequency-domain decision-feedback receiver for MC-CDMA schemes. In *IEEE VTC'05 (Spring)*, volume 1, pages 271–275, Stockholm, Sweden, May 2005.

[50] R. Dinis, P. Silva, and A. Gusmão. IB-DFE receiver techniques for CP-assisted block transmission within DS-CDMA and MC-CDMA systems. In *ISCTA'05*, Ambleside, UK, July 2005.

[51] R. Dinis, P. Silva, and A. Gusmão. IB-DFE receivers with space diversity for CP-assisted DS-CDMA and MC-CDMA systems. *Eur. Trans. Telecommun. (ETT), John Wiley & Sons*, 18:791–802, June 2007.

[52] M. Tüchler and J. Hagenauer. Turbo equalization using frequency domain equalizers. In *38th Annual Allerton Conference on Communication*, pages 1234–1243, Monticello, Illinois, Oct. 2000.

[53] M. Tüchler and J. Hagenauer. Linear time and frequency domain turbo equalization. In *IEEE VTC'01 (Fall)*, volume 4, pages 2773–2777, Atlantic City, USA, Oct. 2001.

[54] B. Vucetic and J. Yuan. *Turbo Codes: Principles and Applications*. Kluwer Academic Publ., 2002.

[55] P. Silva and R. Dinis. SIC and PIC multiuser detection for prefix-assisted DS-CDMA systems. In *IASTED CSA'05*, pages 12–15, Banff, Canada, July 2005.

[56] P. Silva and R. Dinis. Multiuser detection for the uplink of prefix-assisted DS-CDMA systems employing multiple transmit and receive antennas. In *IEEE VTC'06 (Fall)*, pages 1–5, Montreal, Sep. 2006.

[57] P. Silva and R. Dinis. Frequency-domain multiuser detection for CP-assisted DS-CDMA signals. In *IEEE VTC'06 (Spring)*, volume 5, pages 2103–2108, Melbourne, May 2006.

[58] R. Dinis and P. Silva. A frequency-domain receiver for asynchronous systems employing CP-assisted DS-CDMA schemes. In *IASTED SIP'06*, Honolulu, Hawaii, Aug. 2006.

[59] P. Silva and R. Dinis. Turbo detection for the uplink of CP-assisted DS-CDMA systems. In *IEEE PACRIM 2007*, pages 170–173, Victoria, Canada, Aug. 2007.

[60] K. Baum, T. Thomas, F. Vook, and V. Nangia. Cyclic-prefix CDMA: An improved transmission method for broadband DS-CDMA cellular systems. In *IEEE WCNC*, volume 1, pages 183–188, Orlando, USA, Mar. 2002.

[61] J. Shen and A. G. Burr. Iterative multi-user-antenna detector for MIMO CDMA employing space-time turbo codes. In *IEEE GLOBECOM'02*, volume 1, pages 419–423, Taipei, Taiwan, Nov. 2002.

[62] S. Verdú. Minimum probability of error for asynchronous Gaussian multiple-access channels. *IEEE Trans. Inf. Theory*, 32(1):85–96, Jan. 1986.

[63] S. Verdú. Computational complexity of optimum multiuser detection. *Algorithmica*, 4(1):303–312, Dec. 1989.

[64] K. Fazel and S. Kaiser. *Multi-carrier & Spread Spectrum Systems*. Wiley & Sons, 2003.

[65] T. Araújo and R. Dinis. Iterative equalization and carrier synchronization for single-carrier transmission over severe time-dispersive channels. In *IEEE GLOBECOM'04*, volume 5, pages 3103–3107, Dallas, Nov. 2004.

[66] R. Dinis, R. Kalbasi, D. Falconer, and A. Banihashemi. Iterative layered space-time receivers for single-carrier transmission over severe time-dispersive channels. *IEEE Commun. Lett.*, 8(9):579–581, Sep. 2004.

[67] G. J. Foschini and M. J. Gans. On limits of wireless communications in fading enviroments when using multiple antennas. *Wireless Personal Communications*, 6(3):315–335, Mar. 1998.

[68] N. Guo and L. Milstein. Uplink performance evaluation of multicode DS/CDMA systems in the presence of nonlinear distortions. *IEEE J. Sel. Areas Commun.*, 18(8):1418–1428, Aug. 2000.

[69] O. Väänänen, J. Vankka, T. Viero, and K. Halonen. Reducing the crest factor of a CDMA downlink signal by adding unused channelization codes. *IEEE Commun. Lett.*, 6(10):443–445, Oct. 2002.

[70] R. Dinis and P. Silva. Analytical evaluation of nonlinear effects in MC-CDMA signals. In *IASTED CSA'05*, pages 6–11, Banff, Canada, July 2005.

[71] R. Dinis, P. Silva, and T. Araújo. Joint turbo equalization and cancelation of nonlinear distortion effects in MC-CDMA signals. In *IASTED SIP'06*, Honolulu, Hawaii, Aug. 2006.

[72] P. Silva and R. Dinis. Joint multiuser detection and cancelation of nonlinear distortion effects for the uplink of MC-CDMA systems. In *IEEE PIMRC'06*, pages 1–5, Helsinki, Sep. 2006.

[73] R. Dinis and P. Silva. An iterative detection technique for DS-CDMA signals with strong nonlinear distortion effects. In *IEEE VTC'06 (Fall)*, pages 1–5, Montreal, Sep. 2006.

[74] R. Dinis, P. Silva, and T. Araújo. Turbo equalization with cancelation of nonlinear distortion effects for CP-assisted and zero-padded MC-CDMA signals. In *IEEE VTC'07 (Fall)*, pages 1–6, Baltimore, EUA, Sep. 2007.

[75] P. Silva and R. Dinis. Turbo multiuser detection for MC-CDMA signals with strongly nonlinear transmitters. In *IEEE ISCIT 2007*, Sydney, Australia, Oct. 2007.

[76] P. Silva and R. Dinis. A turbo SDMA receiver for strongly nonlinearly distorted MC-CDMA signals. *Can. J. Elect. Comput. Eng. (CJECE)*, 33(1):39–44, Feb. 2008.

[77] R. Dinis, P. Silva, and T. Araújo. Turbo equalization with cancelation of nonlinear distortion for CP-assisted and zero-padded MC-CDMA schemes. *IEEE Trans. Commun.*, 57(7):2185–2189, Aug. 2009.

[78] P. Silva and R. Dinis. Joint turbo equalization and multiuser detection of MC-CDMA signals with low envelope fluctuations. *IEEE Trans. Veh. Technol.*, 58(5):2288–2298, June 2009.

[79] R. Dinis and P. Silva. Iterative detection of multicode DS-CDMA signals with strong nonlinear distortion effects. *IEEE Trans. Veh. Technol.*, 58(8):4169–4181, Oct. 2009.

[80] R. Dinis and P. Silva. Iterative detection of multicode DS-CDMA signals with strongly nonlinear transmitters. In *IEEE ICCCN 2009*, pages 1–6, San Franscisco, USA, Aug. 2009.

[81] R. Dinis and A. Gusmão. A class of signal processing algorithms for good power/bandwidth tradeoffs with OFDM transmission. In *IEEE ISIT 2000*, page 216, Sorrento, Italy, June 2000.

[82] R. Dinis and A. Gusmão. Signal processing schemes for power/bandwidth efficient OFDM transmission with conventional or LINC transmitter structures. In *IEEE ICC'01*, volume 4, pages 1021–1027, Helsinki, June 2001.

[83] A. Brajal and A. Chouly. Compensation of nonlinear distortion for orthogonal multicarrier schemes using predistortion. In *IEEE GLOBECOM'94*, volume 3, pages 1909–1914, San Francisco, CA, Nov. 1994.

[84] A. Saleh. Frequency-independent and frequency-dependent nonlinear models of TWT amplifiers. *IEEE Trans. Commun.*, 29(11):1715–1720, Nov. 1981.

[85] R. Dinis and A. Gusmão. Performance evaluation of OFDM transmission with conventional and two-branch combining power amplification schemes. In *IEEE GLOBECOM'96*, volume 1, pages 734–739, London, Nov. 1996.

[86] O. Väänänen, J. Vankka, T. Viero, and K. Halonen. Effect of clipping in wideband CDMA system and simple algorithm for peak windowing. In *World Wireless Congress*, pages 614–618, San Francisco, USA, May. 2002.

[87] A. Palhau and R. Dinis. Performance evaluation of signal processing schemes for reducing the envelope fluctuations of CDMA signals. In *IEEE GLOBECOM'03*, volume 6, pages 3392–3396, San Francisco, USA, Dec. 2003.

[88] R. Dinis and A. Palhau. A class of signal processing schemes for reducing the envelope fluctuations of CDMA signals. *IEEE Trans. Commun.*, 53:882–889, May 2005.

[89] J. Armstrong. New OFDM peak-to-average power reduction scheme. In *IEEE VTC'01 (Spring)*, volume 1, pages 756–760, Rhodes, Greece, May 2001.

[90] R. Dinis and A. Gusmão. Performance evaluation of an iterative PMEPR-reducing technique for OFDM transmission. In *GLOBECOM'03*, volume 1, pages 20–24, San Francisco, USA, Dec. 2003.

[91] T. Cover. Broadcast channels. *IEEE Trans. Inf. Theory*, 18(1):2–14, Jan. 1972.

[92] K. Ramchandran, A. Ortega, K. Uz, and M. Vetterli. Multiresolution broadcast for digital HDTV using joint source/channel coding. *IEEE J. Select. Areas Commun.*, 11(1):6–23, Jan. 1993.

[93] J. Tellado, L. Hoo, and J. Cioffi. Maximum likelihood detection of nonlinearly distorted multicarrier symbols by iterative decoding. *IEEE Trans. Commun.*, 51(2):218–228, Feb. 2003.

[94] A. Gusmão and R. Dinis. Iterative receiver techniques for cancelation of deliberate nonlinear distortion in OFDM-type transmission. In *Int. OFDM Workshop'04*, Dresden, Sep. 2004.

[95] H. Rowe. Memoryless nonlinearities with Gaussian input: Elementary results. *Bell System Tech. Journal*, 61, Sep. 1982.

[96] G. Stette. Caculation of intermodulation from a single carrier amplitude characteristic. *IEEE Trans. Commun.*, 22(3):319–323, Mar. 1974.

[97] R. Dinis and A. Gusmão. On the performance evaluation of OFDM transmission using clipping techniques. In *IEEE VTC'99 (Fall)*, volume 5, pages 2923–2928, Amsterdam, Sep. 1999.

[98] S. Barbarossa and F. Cerquetti. Simple space-time coded SS-CDMA systems capable of perfect MUI/ISI elimination. *IEEE Commun. Lett.*, 5(12):471–473, Dec. 2001.

[99] N. Benvenuto and S. Tomasin. Iterative design and detection of a DFE in the frequency domain. *IEEE Trans. Commun.*, 53(11):1867–1875, Nov. 2005.

[100] A. Gusmão, P. Torres, R. Dinis, and N. Esteves. A turbo FDE technique for reduced-CP SC-based block transmission systems. *IEEE Trans. Commun.*, 55(1):16–20, Jan. 2007.

[101] M. Mozaffaripour, N. Neda, and R. Tafazolli. Partial parallel interference cancellation and its modified adaptive implementation for MC-CDMA system. In *IEE 3G Mobile Commun. Technol.*, pages 281–285, 2004.

[102] M. Abramowitz and I. Stegun. *Handbook of Mathematical Functions*. Dover Publications, New York, 1972.

# Index

229

# About the Authors

**Paulo Silva** received the Ph.D. degree from Instituto Superior Técnico (IST), Technical University of Lisbon, Portugal, in 2010. He is a Professor at ISE-UAlg (Instituto Superior de Engenharia da Universidade do Algarve). He was a researcher at CAPS/IST (Centro de Análise e Processamento de Sinais) from 1996 to 2005; from 2005 to 2008 he was a researcher at ISR/IST (Instituto de Sistemas e Robótica); in 2009 he joined the research center IT (Instituto de Telecomicações). He has been involved in several research projects in the broadband wireless communications area and he is the author/co-author of several dozens of papers. His main research interests include spread spectrum and equalization techniques.

**Rui Dinis** received the Ph.D. degree from Instituto Superior Técnico (IST), Technical University of Lisbon, Portugal, in 2001. From 2001 to 2008 he was a Professor at IST. Since 2008 he is teaching at FCT-UNL (Faculdade de Ciências e Tecnologia da Universidade Nova de Lisboa). He was a researcher at CAPS/IST (Centro de Análise e Processamento de Sinais) from 1992 to 2005; from 2005 to 2008 he was a researcher at ISR/IST (Instituto de Sistemas e Robótica); in 2009 he joined the research center IT (Instituto de Telecomicações). He has been involved in several research projects in the broadband wireless communications area and he is the author/co-author of over 200 papers published in major international conference proceedings and journals. His main research interests include modulation, equalization and channel coding.